Designing and Building Mini and Micro Hydropower Schemes

Praise for the book...

'This book's strength is that it is based on years of experience out in the field of designing micro hydro systems that work.'

*Dr Arthur Williams, School of Electrical Electronic Engineering,
The University of Nottingham, UK*

'For remote communities lucky enough to live near hill streams or rivers, micro-hydro power is the most cost effective way of generating electricity. And it is clean energy. But it takes years of experience and skill to design the weirs, canals and spillways that are needed. Experienced practitioners take you through the whole design process, with drawings and calculations, so that anyone with good practical building skills can learn enough from the many years of knowledge crammed into this instruction book to build a solid scheme, without over-spending.'

*Ray Holland, Manager, EU Energy Initiative,
Partnership Dialogue Facility*

Designing and Building Mini and Micro Hydropower Schemes

A Practical Guide

Luis Rodríguez and Teodoro Sánchez

Practical Action Publishing Ltd
27a Albert Street, Rugby, CV21 2SG, Warwickshire, UK
www.practicalactionpublishing.org

© Practical Action Publishing, 2011

First published 2011

ISBN 978 1 85339 646 5

All rights reserved. No part of this publication may be reprinted or reproduced or utilized in any form or by any electronic, mechanical, or other means, now known or hereafter invented, including photocopying and recording, or in any information storage or retrieval system, without the written permission of the publishers.

A catalogue record for this book is available from the British Library.

The authors have asserted their rights under the Copyright Designs and Patents Act 1988 to be identified as authors of this work.

Since 1974, Practical Action Publishing has published and disseminated books and information in support of international development work throughout the world. Practical Action Publishing (formerly Intermediate Technology Publications and ITDG Publishing) is a trading name of Practical Action Publishing Ltd (Company Reg. No. 1159018), the wholly owned publishing company of Practical Action Ltd (working name Practical Action). Practical Action Publishing trades only in support of its parent charity objectives and any profits are covenanted back to Practical Action (Charity Reg. No. 247257, Group VAT Registration No. 880 9924 76).

Cover design by Practical Action Publishing
Typeset by S.J.I. Services

Contents

Figures, tables and photographs	ix
Preface	xxi

1	Introduction	1
	Rural electrification problems	1
	Development and promotion of renewable energies in developing countries	3
	Hydropower and its social development potential	4
	Designing a project for a hydroelectric plant	7
	The hydroelectric plant	13

2	Intake	15
	Intake weir with stop planks	18
	Tyrolean type intake	22
	Design of the main components	27
	Intake construction procedures	32
	Technical specifications	40

3	Headrace channel	45
	Channel	45
	Spillway	49

4	Coarse settling basin	53
	Design	54
	Construction procedure	55
	Technical specifications	56

5	Conveyance channel	59
	Design of the channel	61
	Conveyance channel construction process	67
	Technical specifications	76
	Works to complement the conveyance channel	77

6	Silt basin	85
	Description and characteristics of the silt basin elements	85
	Design of the silt basin and complementary works	88
	Silt basin construction process	89
	Technical specifications	96

7	Forebay tank	97
	Design of the forebay tank	98
	Construction process	105
	Technical specifications	105
8	Penstock	109
	Penstock materials	109
	Penstock design	113
	Supports	121
	Anchors	126
	Construction process	135
	Repairing PVC penstock	141
	Technical specifications	142
9	Electromechanical equipment	149
	Hydropower turbines	149
	Hydropower turbine sizing	158
	Electric generators	165
	Control of hydroelectric plants	168
	Examples of the selection of electromechanical equipment	168
	Electricity transmission and distribution	172
10	Powerhouse building	173
	Design	173
	Technical specifications	180
	Procedure for building the powerhouse	183
11	Foundation for the electromechanical equipment	189
	Foundation for the horizontal shaft Pelton turbine	189
	Foundation for the vertical shaft Pelton turbine	197
	Foundation for the Michell-Banki turbine	207
	Foundation for the axial turbine with a vertical shaft	209
12	Assembly of the turbine and generator	219
	Assembly of the horizontal shaft Pelton turbine with two jets	219
	Assembly of the vertical shaft Pelton turbine with three jets	225
	Assembly of the Michell-Banki turbine	228
	Assembly of the axial turbine	230
	Cable connections	232
13	Tailrace channel	235
	Design of the tailrace channel	235
	Construction process	236
	Technical specifications	237
14	Commissioning	239
	Hydraulic tests on civil works	239

Protocol for operating tests of the electromechanical equipment in the powerhouse	243
Acceptance certificate	244

15 Staff training — 247
Basic knowledge — 247
Training in the operation of the hydroelectric system — 248
Basic preventive maintenance of the hydroelectric power system — 248
Additional specialized training — 248

References and bibliography — 249

Annexes
Example 1 Calculation of a mixed intake — 251
Example 2 Calculation of a Tyrolean type intake — 277
Example 3 Calculation of the spillway for the headrace channel — 282
Example 4 Calculations for a coarse settling basin — 284
Example 5 Calculatons for an open channel — 288
Example 6 Calculation of an enclosed channel (pipe channel) — 292
Example 7 Calculations for the silt basin — 293
Example 8 Calculations for the forebay tank — 297
Example 9 Calculations for the PVC penstock — 301
Example 10 Design of the RU PVC and steel pipe — 308
Example 11 Calculation of props or supports — 316
Example 12 Calculations for the anchor — 326
Example 13 Calculation of the foundation of a horizontal shaft Pelton turbine and generator — 345
Example 14 Calculation of the foundation for a vertical shaft Pelton turbine and generator — 349
Example 15 Non-contact hydraulic seals for small micro and mini hydro turbines — 354

Figures, tables and photographs

Figures

1.1	Cost per installed kW for a sample of 15 micro hydropower plants of less than 100 kW	6
1.2	Load diagram of mini hydro power plant, Pozuzo, Peru, 1997	8
1.3	Typical flow duration curve	11
1.4	Components of micro and mini hydropower schemes	14
2.1	Conventional type of intake	15
2.2	Components of an intake weir with stop planks	18
2.3	Profile of the guiding wall	19
2.4	Front view and profile of the intake aperture	20
2.5	Weir, floor slab and stonework apron	21
2.6	Stonework apron with mortar of cement and fine sand	22
2.7	Tyrolean type intake weir	23
2.8	Intake screen (trash rack)	24
2.9	Weir and collecting channel	25
2.10	Tyrolean type intake weir, headrace channel and settling basin	26
2.11	Examples of cross-sections	27
2.12	Diagrams of the forces and pressures acting on the wall	27
2.13	Planning of intake structures	31
2.14	Layout of central lines and levels in the intake	33
2.15	Temporary channelling of the river and excavation of the footing pit	34
2.16	Excavation and casting of the wall footing	35
2.17	Layout of the footing to form the wall body and the intake aperture	36
2.18	Forming of guiding wall and intake aperture	36
2.19	Layout of column footings for excavation purposes	37
2.20	Plinth, reinforced steel in footing and column, determination of foundation levels and laying of concrete	38
2.21	Fixed weir (footing, column, bracing slab and cover slab)	39
3.1	Hydraulic section of the headrace channel	45

3.2	Layout and construction of the platform for the headrace channel	47
3.3	Excavation of the channel	48
3.4	Spillway plan view and elevation	49
3.5	Spillway in elevation	50
4.1	Settling basin plan view and elevation	53
4.2	Diagram of a settling basin	55
4.3	Assembly of prepared steel structure for the walls and floor of the settling basin	57
5.1	Open sections	59
5.2	Closed sections	59
5.3	Geometric elements of the channel section	60
5.4	Hydraulic characteristics of the channel	60
5.5	Maximum hydraulic efficiency channels	64
5.6	Manufacturer's recommendations for using piping as a channel	66
5.7	Open platform	67
5.8	Layout of the centre line of the channel on the platform	68
5.9	Marking points to draw curved border lines for the channel	69
5.10	Marking points to draw curved border lines for the channel	69
5.11	Marking points to draw curved border lines for the channel	70
5.12	Layout of the borders of the channel for the excavation, on the platform	71
5.13	Excavation of the first and second parts of the trapezoidal ditch and moving points P, Q, R and S on the markers	71
5.14	Details of the wooden frame	73
5.15	Woodwork	73
5.16	Alignment, levelling, plumbing and fastening of master frames	74
5.17	Alignment, levelling, plumbing and fastening of master frames in a semicircular section	76
5.18	Aqueduct pipe supported at the ends and suspended in the middle	79
5.19	Balcony type canal	80
5.20	Storm aqueduct	81
5.21	Pressure-breaking drain	82
5.22	Stone wedging for the platform of the conveyance channel	82
5.23	Diversion channels	84
6.1	Silt basin elements	86
6.2	Vertical section view of the silting tank	87
6.3	Silt basin with PVC drain pipe and desilting gate	89
6.4	Extension of the centre line of the channel on the prepared platform	90

6.5	Layout of the silt basin for the excavation and for placing level reference points	91
6.6	Level reference points for the excavation of the silt basin tank	91
6.7	First part of the excavation of the silt basin	92
6.8	Location of the foundation levels, where the ramps on the floor of the silt basin change direction (the excavation for floor slabs is optional, depending on the quality of the soil)	92
6.9	Total excavation of the silt basin tank	93
6.10	Assembly of the steel reinforcement resting on concrete blocks in the silt basin	93
6.11	Layout of the centre line of the silt basin on the floor and forming guidelines for the walls	94
6.12	Homemade PVC elbow pipe for draining the silt basin	96
7.1	Components of the forebay tank and silt basin	99
7.2	Forebay tank elements and details	100
8.1	FU PVC piping and accessories, standard ISO 4422	110
8.2	Steel pressure penstock	111
8.3	PVC pressure penstock	112
8.4	HDPE pressure penstock	112
8.5	Moody diagram to find the friction factor on the inner wall of the penstock	115
8.6	Head loss coefficients due to turbulence	117
8.7	Location of supports between anchors, in a section of the penstock	121
8.8	Types of support profile for steel pipes	122
8.9	Diagram of acting forces when the pipes expand	122
8.10	Diagram of the forces caused by the actual weight of the concrete block	123
8.11	Forces involved in the calculation of the support	123
8.12	Section of a penstock showing internal and external diameter	125
8.13	Cross section of the penstock	126
8.14	Types of anchor for different landscape profiles	127
8.15	Weight of the penstock and the water (F1)	128
8.16	Frictional force (F2) and expansion joint	128
8.17	Force due to the hydraulic pressure on the anchor (F3)	129
8.18	Weight of penstock acting parallel to the penstock (F4)	129
8.19	Frictional forces in the expansion joint (F6)	130
8.20	Stress caused by the hydrostatic pressure within the expansion joint (F7)	130
8.21	Stress caused by the change in direction of the water moving in the elbow pipe or bend	131

8.22	Hydrostatic forces due to the change of section of the penstock (F9)	131
8.23	Acting forces for calculating Anchor 1, Case I, when the pipe expands	134
8.24	Positioning of stakes and levels in the centre line of the piping drawn on the ground	135
8.25	Excavation depth, h_i, based on each level fixed on the ground	136
8.26	Penstock joints	137
8.27	Fitting the penstock in the forebay tank with the vent pipe	138
8.28	Fixing of the curve while laying the concrete	139
8.29	Dimensions and covering of the ditch	139
8.30	Pipe repair steps	143
9.1	Graph for the selection of the type of turbine for a MHS	153
9.2	Different types of turbine (clockwise, from top left): Pelton, Pelton, Francis, Francis semiaxial, Kaplan, Kaplan, Francis	154
9.3	Efficiency of small hydraulic turbines with partial flows	157
9.4	Geometry of the Michell-Banki runner	161
9.5	Efficiency of the Michell-Banki turbine at partial flows	163
9.6	Francis turbine	165
9.7	Basic dimensions of a Francis turbine	166
10.1	Turbine room layout and powerhouse dimensions	174
10.2	Powerhouse plan view	175
10.3	Front elevation and cross-section of the powerhouse	176
10.4	Plan view of foundations, footings and details	177
10.5	Wooden roof structure	178
10.6	Electrical installation in the powerhouse (plan view)	179
10.7	Putting in survey poles, drawing centre lines and foundations	184
10.8	Alignment of continuous footings, using survey pole markings	185
11.1	Centre lines of the turbine and generator with respect to the centre line of the penstock	189
11.2	Locaton of the horizontal shaft Pelton turbine with respect to the rest of the components	190
11.3	Sizing of the foundation, based on the measurements of the metal base of the turbine	191
11.4	Layout of the centre line of the pipe on the floor and the wall	193
11.5	Profile of the layout of a horizontal shaft Pelton turbine	193
11.6	Marking the foundation's centre lines	194
11.7	Layout for digging the foundation pit for the turbine and generator	194
11.8	Foundation pit for the base of the turbine and generator	195

11.9	Fitting the steel reinforcement cage onto the concrete blocks, prior to the formwork	195
11.10	Metal base for the turbine and generator, anchored to the reinforced concrete foundation base	196
11.11	Plan view showing the location of the turbine and generator shafts with respect to the centre line of the penstock	199
11.12	Layout of electromechanical equipment with a vertical shaft Pelton turbine, in accordance with the dimensions of the equipment	199
11.13	Layout of the centre line of the pipe on the ground and on the wall	200
11.14	Side view of the casing of a vertical shaft turbine and the base of the generator	201
11.15	Centre line of the penstock (AB), centre line perpendicular to it (CE), and parallel lines that go through the vertices of both the casing and the base of the generator	202
11.16	Removal of pre-assembled equipment and layout of the casing and the base of the generator on the ground	202
11.17	Layout of the base for the turbine and the generator and auxilliary components, for excavation purposes	203
11.18	Foundation pit	203
11.19	Steel reinforcement cage for the foundations of the vertical shaft Pelton turbine and the generator	204
11.20	Levelling of the casing and the metal base of the generator; forming and cementing the concrete foundation	205
11.21	Casing of the vertical shaft turbine and metal base of the generator, anchored to their concrete foundations	205
11.22	Dimensions and reinforcement of the buffer column	206
11.23	Location of the centre lines of the turbine and generator with respect to the centre line of the penstock	207
11.24	Layout of electromechanical equipment with a Michell-Banki turbine, similar to that of the horizontal shaft Pelton turbine	208
11.25	Levelling, alignment and separation joint between the Michell-Banki turbine and the nipple joined to the valve, before pouring the concrete	209
11.26	Layout of electromechanical equipment with a low head turbine (Kaplan propellor type), Las Juntas MHS, Peru	210
11.27	Sizing of foundations (1, 2 & 3) for the low head turbine and generator	211
11.28	Steel layout (on the base of the generator and base of the turbine)	213
11.29	Layout of the centre line of the penstock	214
11.30	Casing presentation and layout of centre lines	214
11.31	Layout of the protecting wall for the draft tube	215

11.32	Construction of the protecting wall for the draft tube and layout of the foundations	215
11.33	Layout of the ditch for the generator base, walls for the draft tube and casing supports	216
11.34	Foundations for the turbine and generator	217
12.1	FU and RU PVC accessories	219
12.2	Layout of the turbine components	220
12.3	Assembly of the turbine components	221
12.4	Coupling the valve, nipple, casing and nozzle on the metal base	222
12.5	Continuation of the assembly of the horizontal shaft Pelton turbine components	224
12.6	Culmination of the assembly of the horizontal shaft Pelton turbine and generator	224
12.7	Pieces of the vertical shaft Pelton turbine and metal base for the generator	225
12.8	First stage of the assembly of a vertical shaft Pelton turbine and metal generator base	226
12.9	Stage two of the assembly of the vertical shaft Pelton turbine and generator	228
12.10	Assembly of the Michell-Banki turbine and accessories for coupling it to the penstock	229
12.11	Stage one of the assembly of a low head turbine	230
12.12	Final assembly of a low head turbine and generator	233
12.13	Earthing system	234
13.1	Location of the tailrace channel (in elevation) for Pelton or Michell-Banki type turbines	235
13.2	Profile of the tailrace channel for an axial turbine	236
14.1	Example of an acceptance certificate	246
A1.1	Cross-section of the river for the intake design	251
A1.2	Preliminary sizing of the wall	252
A1.3	Analysis of intervening forces	253
A1.4	Final resizing of the wall	255
A1.5	Side view and front view of the window ledge and dimensions of the intake aperture	257
A1.6	Dimensions L and t in the intake aperture	258
A1.7	Section of the headrace channel (in normal conditions)	259
A1.8	Water level and hydraulic characteristics in the intake aperture and the headrace channel during the rainy season	260
A1.9	Section of the headrace channel (in high flow conditions)	262
A1.10	Geometric characteristics of the column with stop plank grooves	263
A1.11	Dimensions of the column with stop plank grooves	264

FIGURES, TABLES AND PHOTOGRAPHS

A1.12	Steel distribution in column and footing	265
A1.13	Geometric characteristics of the bracing slab	266
A1.14	Sizing of the bracing slab	267
A1.15	Steel distribution in the bracing slab	268
A1.16	Characteristics of the stonework	269
A1.17	Dimensions of the plank	270
A1.18	Wood distortion caused by drying	271
A1.19	Removable weir made of wooden planks	272
A1.20	Diagram of the bending moment and shearing force for a beam embedded at both ends	273
A1.21	Bending moments over the slab due to the water thrust	274
A2.1	Outline of intake	277
A2.2	Distance between screen elements	278
A2.3	Inner dimensions of the collecting channel	280
A2.4	Dimensions at the beginning of the cross-section of the collecting channel. The height increases gradually, in accordance with the established slope	281
A2.5	Distribution of the reinforcement in the collecting channel and the concrete	281
A3.1	Design of the spillway	282
A3.2	Final design of the spillway and headrace channel	283
A4.1	Inner dimensions of the settling basin	285
A4.2	Distribution of the steel structure in the settling basin	286
A4.3	Measurement of the outflow through the bottom gate	287
A5.1	Calculation of a conveyance channel with open sections of a maximum hydraulic efficiency	290
A5.2	Final dimensions of the rectangular, trapezoidal and semicircular sections of a channel, to deliver 0.600 m^3/s	291
A7.1	Design of the settling basin	294
A7.2	Design of complementary works	295
A7.3	Distribution of reinforcement steel in the settling basin	296
A8.1	Determining the value of m	297
A8.2	Final sizing of the forebay tank components joined to the settling basin	298
A8.3	Details of the screen and bracket	299
A8.4	Steel distribution in the forebay tank	300
A9.1	Stake out and registration of distances and angles to determine the ground profile	301
A9.2	Determining the profile of the ground and the penstock	302
A10.1	Profile of the penstock with steel pipes in the lower part and PVC pipes in the higher part	308

A11.1	Preliminary sizing of the support block	316
A11.2	Support profile and preliminary sizing	317
A11.3	View of a section of penstock	317
A11.4	Breakdown of the weights of the concrete block and of acting forces	319
A11.5	Areas and lever arms to determine X_G of the block	320
A11.6	Resulting forces for calculating the stability of the support	321
A11.7	Breakdown of the weights of the concrete block and change of direction of the force of friction (Case II)	322
A11.8	Resulting forces for calculating the stability of the support when the pipe contracts	324
A12.1	Field data and data from steel pipe catalogues to calculate the anchor for the lower part	326
A12.2	Preliminary sizing of the concrete anchor	327
A12.3	Acting forces for calculating anchor 4, when the pipe expands	328
A12.4	Calculating the weight of the concrete anchor	329
A12.5	Diagram of concrete weights and acting forces	329
A12.6	Diagram of areas	330
A12.7	Diagram of resulting forces for calculating the soil stability	331
A12.8	Acting forces for calculating the anchor when the pipe contracts	332
A12.9	Concrete weights and acting forces when the pipe contracts	333
A12.10	Diagram of resulting forces for calculating the soil stability (Case II)	334
A12.11	Profile of the penstock with steel pipes in the lower part and PVC pipes in the higher part	336
A12.12	Preliminary sizing in perspective and profile of the outward facing anchor	337
A12.13	Diagram of partial weights of the concrete anchor and acting forces (Case I, when the pipe expands)	338
A12.14	Diagram of partial weights of the concrete anchor and acting forces (Case I, when the pipe expands)	339
A12.15	Diagram of partial weights of the concrete anchor and acting forces (Case II, when the pipe contracts)	340
A12.16	Diagram of partial weights of the concrete anchor and acting forces (Case II, when the pipe contracts)	341
A13.1	Sizing of the foundation bases for the turbine and generator	346
A13.2	Distribution of steel structure	348
A14.1	Pre-assembly of the vertical shaft Pelton turbine and the metal base of the generator	350
A14.2	Sizing of the foundation base for the vertical shaft Pelton turbine and generator	351

A14.3	Steel structure in the foundation base for the vertical shaft Pelton turbine and generator	353
A15.1	Boundary layer	355
A15.2	Position of the hydraulic seal	356
A15.3	Dimensions of the hydraulic seal	357

Tables

1.1	Use and simultaneity factors	10
1.2	Typical k values	12
1.3	Classification according to power range	12
2.1	Specific soil weights and friction angles	28
2.2	Types, admissible forces and friction coefficients of foundation ground	29
2.3	Density of materials in kg/m^3 (National Construction Regulations)	40
2.4	Contributing coefficients for 'in volume' mixes, with 5% waste per m^3 of concrete	41
2.5	Contributing coefficients for 'in volume' mixes, with 5% waste per m^3 of concrete	41
2.6	Contributing coefficients for 'in volume' mixes, with 5% waste per m^3 of concrete	41
2.7	Contributing coefficients for sand-cement mortar, in volume per m^3 (including 5% waste)	42
2.8	Contributing coefficients for primary plastering with cement and sand mortar, per m^2 (including 5% waste)	42
2.9	Contributing coefficients for whitewashing with pure gypsum per m^2 (including 5% waste)	42
2.10	Contributing coefficients for concrete floors, pavements or floor plates per m^2 (including 5% waste)	43
2.11	Resistant concrete floors of cement, sand and stone, per m^2 (including 5% waste)	43
4.1	Settling speed of the particles, depending on the diameters	54
5.1	Recommended wall slope for trapezoidal section channels	62
5.2	Maximum flow velocity recommended	62
5.3	Manning coefficients	63
6.1	Recommended speeds to settle different sizes of particles	87
7.1	Range of flows for different internal diameters of penstock	101
7.2	Calculations for the components of the channel, silt basin and forebay tank	102
7.3	Distribution of steel reinforcement in the silt basin and forebay tank	103

7.4	Dimensions for the different components of a forebay tank according to the flow (m^3/s) of the MHS	104
8.1	Surface roughness values k (mm)	115
8.2	Summary of the forces intervening in the calculation of the anchor (see Figure 8.23)	132
8.3	Socket joint insertion lengths	137
8.4	Materials used for repairing PVC pipes	141
8.5	Rieber union penstock systems, standard NPT ISO 4422	146
8.6	Technical characteristics of PVC	147
8.7	Rigid PVC pipes	147
8.8	Steel piping systems	148
9.1	Hydraulic turbines and specific speeds (SI)	155
9.2	Main characteristics of hydraulic turbines	156
9.3	Influence of the number of poles in the frequency	160
9.4	Values of K_a, depending on the admission angle ($\alpha_1 = 16°$)	164
11.1	Characteristics of the equipment supplied to Las Juntas MHS	212
A1.1	Results obtained	260
A1.2	Results obtained	261
A1.3	Mechanical characteristics of the timber frequently used in Peru	273
A2.1	Results from calculations	279
A5.1	Parameters to be calculated	288
A5.2	Parameters to be calculated	288
A5.3	Parameters to be calculated	289
A5.4	Recommended values for B_L according to flow	290
A5.5	Thickness of simple concrete channels	291
A5.6	Some measurements of the calculated sections as indices for determining the most economic section of the channel to deliver 0.600 m^3/s	291
A9.1	Matching head of penstock to commercial PVC pipes according to class	304
A9.2	Data processing for different sections of the penstock	305
A9.3	Calculation of friction losses for different sections	305
A9.4	Calculations of speed for different sections of penstock	305
A9.5	Calculation of total losses for different sections of the penstock	306
A9.6	Losses as a percentage of the head	306
A9.7	Calculation of values for a	307
A9.8	Pressure increase due to surge	307
A9.9	Total pressure head including surge pressure	307
A9.10	Calculation of the safety factor for each section of the penstock	307
A10.1	Data for commercial PVC pipes	309
A10.2	Data for commercial steel pipes	309

A10.3	Results for the different materials to be used for the different sections of the penstock	309
A10.4	Calculating f	310
A10.5	Calculation of friction losses	310
A10.6	Summary of the calculations	311
A10.7	Total losses in the penstock	311
A10.8	Losses as a percentage of the pressure head	311
A10.9	Calculation of the speed of propagation wave	312
A10.10	Pressure in each section of the penstock	312
A10.11	Total loss in each section of the penstock	312
A10.12	Materials and their characteristics for each section of the penstock	313
A10.13	Summary of the pipes that comply with the penstock conditions	313
A10.14	Sizing of the PVC penstock, Sondor MHS	314
A10.15	Sizing of the PVC and steel penstock, Chetilla MHS	315
A11.1	Weights for the different blocks	319
A11.2	Design of the support for the steel pipe: Chetilla MHS	325
A12.1	Summary of calculations of the stability of concrete blocks	338
A12.2	Design of anchor no. 02 for the PVC pipe, Chetilla MHS	343
A12.3	Design of anchor no. 04 for the steel pipe, Chetilla MHS	344
A13.1	Characteristics of the equipment supplied	345
A14.1	Characteristics of the equipment supplied to the Yanacancha MHS	349

Photographs

2.1	Intake weir functioning with six columns and no guiding walls	16
2.2	Location of an intake weir with stop planks by a spring. A headrace channel was not necessary, but a conveyance channel was	17
3.1	Headrace channel	51
5.1	Pipe to convey water where an open channel is impracticable	65
5.2	Excavating the ditch using a wooden frame	72
5.3	Aqueduct supported by a steel lattice beam	79
6.1	Layout of the silt basin	90
6.2	Placing steel reinforcements in the silt basin	94
7.1	Forebay tank, silt basin and complementary components	106
8.1	Levelling and fitting the penstock at the entrance of the powerhouse, with respect to the level of the finished floor and the location of the turbine	136

8.2	Fitting the repair unions into the damaged PVC penstock	142
10.1	Powerhouse constructed with local materials	187
11.1	Foundation for the turbine-generator unit	198
11.2	Pre-assembly of the casing, inlet pipe and alignment of the centre line of the penstock with the centre line of the inlet pipe in order to draw the layout for the excavation of the foundation for the turbine and generator	201
11.3	Metal base for the generator, anchored to its concrete foundation. The turbine casing is coupled to the draft tube and the metal base of the generator	217
12.1	Application of the silicone seal	222
12.2	Fitting the plumber blocks and roller bearings	223
12.3	Michell-Banki turbine and generator	229
12.4	Axial turbine, showing the nose cone, runner, diffuser section and the lower bearing housing, installed together on the shaft	231
A1.1	Mixed weir showing the guide walls, the fixed weir (columns), and the intake aperture	276

Preface

This book is based on the experience acquired by the authors in hydroelectric schemes that produce less than 500 kW of power. However, the concepts and engineering principles are also applicable to larger schemes; therefore anyone interested in using this book for larger size plants can use their own judgement and evaluations. Unlike other texts that have a bearing on the electromechanical equipment, this one is designed for technicians interested in implementing the whole scheme. Emphasis is therefore placed on the civil works, which usually require larger investments than the electromechanical equipment.

The term civil works relates directly to all the components of the scheme, either because many of them consist entirely of civil works, or because civil works are required to lay their foundations or to house, anchor or support them. Furthermore, the civil works usually require local manpower as well as inputs that are normally available nearby. Therefore, a large proportion of the scheme's components can be built entirely with local resources, based on the theory and procedures described herein, while the electromechanical equipment is being procured. This book is divided into two parts: the main text contains the theory and the referential information required to design, build or select the components of the scheme; the Annexe contains examples of the civil works calculations required to build or install such components.

The main text is divided into 15 chapters. Chapter 1 deals with the rural electrification problem and the proposed alternative to prioritize electrification using renewable energies, particularly with the construction of hydroelectric schemes. It also contains a general description of a typical power plant and its main components. Chapters 2 to 7 describe the components that impound the water from the main source upstream and transport it to where it will start being converted to energy, i.e. the intake, the headrace channel, the coarse settling basin, the conveyance channel, the silt basin and the forebay tank. Chapters 8 and 9 describe the components in which the potential energy of the water is transformed into electricity, i.e. the penstock and the electromechanical equipment, which forms part of the powerhouse. Chapters 10 and 11 refer to the civil works required for the electromechanical equipment, such as the building to house the equipment and the foundation for the turbine and generator. The assembly of the electromechanical equipment is covered in Chapter 12. Chapter 13 describes the tailrace, the component that conveys the residual water back to the main source downstream. Finally, Chapters 14 and 15 refer to matters concerning the start-up of the scheme, such as operating tests and staff training, respectively.

The Annexes comprise 14 sample calculations for the civil works, corresponding to the construction or installation of the scheme's components. Finally, the last annexe deals with the design of hydraulic seals for micro and mini hydro turbines, which I invented while I was the Energy Team Leader in the Practical Action Peru Office. I would therefore like to thank the Director of Practical Action Peru, Alfonso Carrasco, and my colleagues in the Energy Team in Peru from 1995 to 2004, Luis Rodriguez, Saul Ramirez, Rafael Escobar and Gilberto Villanueva, and also Dr Mark Waltham with whom I worked on micro hydro applications when I was seconded to the Practical Action UK office in 1994 and 1995 (at that time ITDG).

CHAPTER 1
Introduction

Rural electrification problems
Poor people's access to energy

Access to energy is a fundamental element for human development; institutions such as the World Bank, United Nations, the European Union, the World Energy Council and others, consider energy to be essential for promoting or improving a number of basic services, such as lighting, safe water, health, education, communications and others (UNDP). There is also a general consensus that the Millennium Development Goals will only be achieved if universal access to energy is achieved first. That includes electricity, a fundamental source of energy.

Nevertheless, achieving universal access to electricity poses a huge challenge, given the magnitude of the electricity deficit – according to the statistics of organizations such as the World Bank and others more than 1.6 billion people in the world have no access to this service (IEA, 2006). On the other hand, despite government efforts in developing countries, the progress made in terms of providing access to electricity has been slow and uneven. There has been a reasonable increase in the electricity coverage in urban areas, thanks to the investments made by both the state and the private sector. In rural areas, however, state investments have been slow or non-existent and private investments are not viable under the current poor and isolated market conditions. Even though approximately 480 million rural dwellers gained access to electricity during the 1970s and 1980s (World Bank, 1996), on a global scale, the number of inhabitants without electricity increased by about 150 million between 1970 and 1990. The progress made so far since 1990 has not been impressive and, in percentage terms, the development has been very limited.

According to recent literature, the energy requirements of poor people are very small, particularly in rural areas, where households rarely exceed between 30 and 50 kWh a month. Field data also reveal that a large proportion of the population consume much less than these figures; for example, in the majority of the communities in the Peruvian Andes, many users require less than 10 kWh a month and between 60 per cent and 70 per cent of the families rarely exceed 20 kWh a month (Sánchez, 2006). The limited amount of energy they use is mainly for basic lighting purposes; the main impact is improvement in children's education, access to information and a minimum of entertainment.

Energy alternatives

The following are the two main energy alternatives for rural populations:
1. The grid for urban fringe sectors and rural communities located near the grid.
2. Small decentralized systems consisting of small hydroelectric schemes, wind generators, solar photovoltaic panels, biogas or others based on renewable sources of energy, as well as small diesel systems.

Unfortunately, the contribution of the grid as a rural electrification alternative is very restricted and becomes increasingly less viable the further the communities are from large towns and/or the routes of medium or low voltage transmission lines. For example, the feasibility studies that the World Bank carried out in Brazil at the beginning of this decade show that a connection to the grid would cost US$6,500 per family for distances of between 10 and 20 km with three phase grids. Needless to say, the price increases the further away the towns to be served are located (World Bank, 2003).

In contrast, the decentralized systems alternative is becoming increasingly more appropriate, given its adaptability to small demand conditions and the fact that the costs tend to decrease as the implementation of the projects progresses and national or local capacities are created.

Barriers to the implementation of small isolated power generating schemes

As can be appreciated, the needs are clear and so are the options. Many studies show that the energy requirements of families with limited financial resources are 'modest', particularly in rural areas. For example, a global estimate by the World Energy Council (2000) concluded that the energy requirements of the 2 billion people with no access to electricity were equivalent to only 7 per cent of the total electricity generated at present. It also suggested that, despite this low figure, attempts to overcome this situation have failed. This conclusion is even more dramatic considering that this failure has occurred at a time in history when all fields of technology, including energy, are developing at a rapid rate.

Estimates of the investments in energy required in order to maintain the current consumption trend for the next three decades, also reach astronomical figures of US$16–17 trillion. Of course there is no doubt that the companies involved in this business will invest as much as is necessary to maintain the energy supply situation and more, if necessary. This gives rise to the following question: Why is it that part of this expected trillion dollar investment will not cover the modest 7 per cent mentioned above and thus overcome the lack of energy services in the world and also contribute to the achievement of the Millennium Development Goals? The reply can be found in the documents of various organizations and individuals working on energy and development. 'There are great barriers' that prevent this process from moving forward. Even though there are differences over what these barriers are, at least there is a consensus that they are related to the social, technological, economic and financial circumstances and the institutions and policies of each particular context and national environment.

The barriers are such that even if enough money was available to implement the infrastructure required to generate all the energy needed, universal access to modern energy is still an enormous challenge. As mentioned above, the barriers have particular connotations in different contexts, yet there are some common patterns that could be generalized to a large extent, for which common solutions or solution frameworks could be suggested (Sánchez, 2006). In view of these particular characteristics, dealing with the situation and eliminating the barriers becomes an even greater challenge.

Development and promotion of renewable energies in developing countries

Since the early 1970s, governments, the civil society and international cooperation agencies have implemented a significant number of projects aimed at promoting access to energy in rural areas, particularly electricity. During the 1970s and 1980s, the efforts were aimed mainly at the development and demonstration of renewable energy technologies, the most prominent examples being technologies for manufacturing and implementing mini and micro hydroelectric schemes, wind pumps and small biogas digesters. Since the 1990s, tremendous efforts have been made to promote micro loans and financial models, in order to encourage private investments so that they can become the driving force behind the delivery of rural electricity services. This emphasis continues to this day, even though more attention is also being paid to integrated projects that include all matters related to the viability of small isolated systems: technology, financing, operation, benefits, complementary services and legal frameworks.

Brief historical account of the use of hydropower

The use of hydropower dates back many years. Small and simple machines such as water wheels were used for grain mills and other activities. Hydropower is currently a very important source of energy in the world, contributing mainly to the generation of electricity to feed national grids. Systems as large as the Itaipu scheme in Brazil, which generates about 13,000 MW, or the Three Gorges in China which generates more than 18,000 MW, demonstrate their capacity to generate large amounts of energy as well as the maturity and importance of the technology used.

Despite the great technological progress, there are still enormous resources in many of the world's countries that remain untapped, as occurs in the majority of Latin American countries. For example, Peru has the potential to produce an estimated 65,000 MW of hydropower, yet so far there is only an installed capacity of 2,500 MW representing about 50 per cent of the total installed capacity and supplying 93 per cent of the electricity consumed in the national grid.

Mankind's activities require some degree of energy or other. It has recently been proved that generating energy based on hydrocarbons will be unsustainable in the medium term, because oil is on the way to depletion and the same will occur with gas and coal within a few dozen years. Moreover, hydrocarbons produce greenhouse gases that are seriously contaminating the environment.

Hydropower is a clean and renewable energy that provides one of the best sources for generating electricity and mechanical energy, particularly in Latin American countries where the demand for rural energy could be covered by small-scale water power plants, with minimum effects on the environment.

In this respect, hydropower can be an important source of clean energy, as long as a respect for the environment and for the different communities in the respective catchments is taken into consideration during the implementation. However, perhaps its most important feature is that it can be used conveniently to generate energy on a small scale and in an isolated manner for rural electrification purposes; this is one of the most appropriate and financially competitive alternatives for providing energy to rural populations.

Hydropower and its social development potential

Technical dossier for the implementation of a small hydroelectric scheme

Hydroelectric projects usually require long implementation periods, both in the pre-investment and investment stages. The installation of a hydroelectric plant usually involves several stages, from pre-feasibility studies to engineering designs. Normally, investment decisions are taken based on feasibility studies, once the actual potential has been identified through accurate power estimates on the one hand and the demand for energy on the other. The construction of a plant may be technically feasible and suitable, but unless there is clearly a market, or better still a buyer for the electricity generated, the decision to set up the scheme cannot be taken.

Once the investment decision is taken, the next step is the final engineering design, which consists of an accurate scaling of each component of the scheme, the size of every sub-component of the civil works, the dimensions of the pipes and powerhouse and detailed specifications of each piece of equipment to be used, including their origin. Well-known calculation methods and the corresponding standards must be employed during this process.

Nevertheless, the technical aspects are not the most important to ensure the success of a hydroelectric scheme; it is the market feasibility and the investment costs required for the implementation that matter. As mentioned previously, long periods are required for the installation of hydroelectric schemes, therefore the recovery is slow. However, since these schemes do not consume fuel, their marginal costs are usually much lower than other energy alternatives; therefore, they are easier to deliver.

Mini, micro and pico hydroelectric schemes are usually aimed at meeting a specific demand for energy in isolated towns where electricity is not normally a profitable business. However, they are a good alternative for this purpose, because they do not require a fuel supply, which could be complicated and unreliable, particularly when there are no good means of communication. Hydroelectric schemes that take advantage of local resources are more familiar for users and easier for local agents to operate and maintain.

Costs of hydroelectric plants

The costs of hydroelectric schemes are usually calculated by installed kW (installed unit cost). As in other production fields, the economy of scale has an influence on this cost; in other words, the larger the scheme (and therefore the more power generating capacity), the lower the unit cost. In the past, 'high unit costs' was one of the reasons for the limited interest shown in the installation of small-scale hydroelectric schemes, particularly micro hydro schemes.

The average implementation estimated at present for large hydroelectric schemes is about US$1,000 per installed kW, whereas for micro hydroelectric plants, depending on the country they are installed in and the origin of the equipment, a wide range of costs have been reported, which are usually high compared with other alternatives such as small diesel generators.

According to the authors' experience, using local manufacturing technologies and local techniques (national consultants), the costs are substantially lower than those reported internationally. Figure 1.1 illustrates a sample of micro hydroelectric plants designed and installed by the authors in Peru between 1995 and 2003, selected at random. As can be appreciated, the average cost is moderate, varying between US$2,000 and US$4,000 in the majority of cases, with an average close to US$3,000.

The costs mentioned above are higher when the system requires civil works (very long canals, difficult and costly intakes, rugged geography, among others) and/or medium voltage grids with high and low current transformers, or when the houses are not grouped together in one place but are semi-dispersed. The other extreme (low costs) occurs when the costs of the civil works are very low, when medium voltage is not required or when the demand corresponds to an application that does not require distribution grids.

In the case of small hydro plants for community electricity services, the contribution of manpower and local materials by users does not reduce costs as substantially as many promoters would like. According to the authors' experience, users generally contribute less than 10 per cent of the total and, in very exceptional cases, 15 per cent. Nevertheless, this participation becomes more interesting from a users' ownership and education approach, as these aspects contribute to the sustainability of community systems.

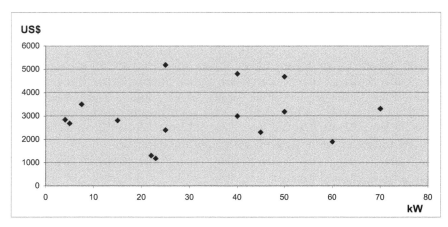

Figure 1.1 Cost per installed kW for a sample of 15 micro hydropower plants of less than 100 kW
Source: Authors

In exceptionally favourable cases, the cost per installed kW can be reduced substantially. For example, when the channel is short, or when the houses are grouped together and located near the powerhouse, among other factors. In such cases, the costs of the civil works, which have the greatest influence on total costs, are reduced to a minimum and it is possible to get costs as low as US$1,000 per installed kW.

The main risks that a hydroelectric plant is exposed to can be summarized as follows:

1. The risk of a shut-down of the hydroelectric system. This can occur for various reasons, the following being the most important:
   - Inadequate estimates of the water resources, i.e. a deficient hydrological study. Therefore the design flow was too high, causing problems in the turbines due to the shortage of water.
   - Inadequate selection of the turbine, e.g. a turbine that is too big for the existing flow, as a result of which it would operate inefficiently and could even come to a standstill. Flaws of this nature (when the turbines stop) occur more easily when reaction (Francis and axial) turbines are used, as impulse turbines have a more even performance, even with relatively low partial flows.
2. Maintenance and/or operating flaws. It is understandable that poor maintenance of the plant as a whole or any of its parts could cause the plant to shut down. For example, a cave-in or obstruction of the canal or any part of the civil engineering work could leave the plant without water and therefore cause it to stop. Likewise, if a bearing or any other mechanical element should break, the plant could also shut down. Stoppages due to mechanical failures can take longer to fix when no

spares are available. Furthermore, operating failures that are also of a technical nature may be due to human error and occur when the operator is not properly trained to operate the entire system correctly.
3. Quality of the equipment. The quality of the equipment depends mainly on the manufacturer's qualifications. For this reason, consideration should be given to the manufacturer's background, the materials used, the manufacturing process and standards used, especially for mini hydro plants.
4. Quality of the installation of the machinery. Particularly for larger than mini hydroelectric plants, the quality of the installation is important. A good assembly implies that the machines must be well aligned with each other, particularly the turbine with the generator and the intermediate components to be used. The suppliers of the machines usually offer a technical service for the installation of the systems. Contracting the manufacturers to install the systems is advisable to a certain point because it prevents arguments between the suppliers and the installers if anything goes wrong with part or all the equipment. It is therefore recommended that for projects larger than mini hydroelectric plants, priority should be given to turnkey contracts.

Designing a project for a hydroelectric plant

Technically speaking, designing a hydroelectric plant involves a group of documents that contain processed calculations, a detailed description of the results, including explanatory tables and other elements to verify the accuracy of the data and methods used. Each section usually contains a document of at least one volume; the majority of the studies require detailed plans (civil works, equipment and grids), therefore each volume also contains the corresponding detailed plans.
1. Electricity market surveys (demand)
2. Hydrological study
3. Civil works: Scaling of the civil works and detailed calculations of the physical dimensions of each of the components mentioned above (intake, headrace channel, silt basin, forebay tank, powerhouse, tailrace), as well as complementary works such as the access road, camp, etc.
4. Electromechanical equipment (including the technical characteristics of each of the machines and auxiliary equipment)
5. Transmission lines (usually medium or high voltage)
6. Distribution grids
7. Geological studies

Estimation of the demand

This comprises a detailed study of the current demand for energy and its potential performance during the useful life of the plant. It can either refer to the energy consumed in a particular town, or to the supply of a national grid.

8 DESIGNING MINI AND MICRO HYDROPOWER SCHEMES

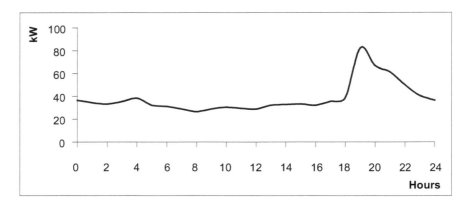

Figure 1.2 Load diagram of mini hydro-power plant, Pozuzo, Peru, 1997

In the case of a town, it is important to determine the present energy consumed in the town, to which end the specific power per family is determined in accordance with national standards, then all other additional loads (demands) are added, such as the industrial and institutional sectors, special demands and public lighting. Once the present energy demand is obtained, the future power is estimated taking the demand growth factors into account.

It must be borne in mind that the energy demands of different sectors do not increase in the same way; therefore one of the ways to estimate them is using growth rates per sector. For example, for domestic demand, the population growth rate can be used, whereas for industrial demand, the historical growth data of this sector should be employed. If these are not available, then an index from another place with similar characteristics can be applied.

The demand is usually estimated for periods of between 20 and 30 years, which is the estimated life span of a hydroelectric plant, although occasionally it is estimated for longer or shorter periods, depending on the case and the designer's experience. The design must cover the supply for all or part of the useful estimated life of the plant. In some cases, for investment reasons, the implementation takes place in stages in accordance with the increase in the demand.

When the power generated goes to the national grid, it is important to determine the present and future needs of this grid and to know the rules for selling the generated power. 'Deregulated' markets, such as those in Chile, Argentina and elsewhere, have a supply system based on the free competition of prices per power unit; in such cases, energy can only be sold if the selling price is competitive with those of other electricity producers. Generally, the marginal costs of electricity produced by hydroelectric plants are very low compared with that produced by thermal plants; therefore hydroelectric plants are more likely to be used to supply baseloads (off-peak loads), whereas thermal plants are more likely to be used to supply peak-load.

It is worth bearing in mind that mini, micro and pico plants designed to supply small, isolated towns rarely reach demands of an 'industrial' scale. In the majority of cases, the demand is mainly for domestic consumption and possibly for small service workshops for the community, such as farm tool repair shops, grain mills, and the conservation of small quantities of food products. Power is used for domestic purposes mainly at night (see Figure 1.2), whereas for institutions and productive purposes the demand is during the day. Therefore, in general these demands should not be added; instead, the behaviour of the different demands should be analysed. In small towns or villages, domestic demand is predominant in most cases; therefore, it would be enough to quantify the domestic demand in such cases. Nevertheless, for methodological reasons, it is better to use the so-called use and simultaneity factors, so as to obtain a more adequate result for the power design (see Table 1.1).

Domestic power consumption in rural areas

There is an abundance of information showing that the domestic power consumption in rural areas is very limited, often averaging between 30 and 50 kWh a month per family. A field study conducted in 14 communities in the Andean and Amazon fringe areas of Peru with access to small hydroelectric plants or small diesel generators (between 20 and 200 kW of power), shows that the maximum average consumption in peak hours rarely reaches 500 W per family (Sánchez, 2006).

Considering all these parameters, Table 1.1 shows some figures for the use and simultaneity factors that the authors worked with, obtaining good results in estimating the demand. The power columns are left blank for the user to fill in for his or her situation. Despite the power per family suggested above, we estimate the demand assuming 300 W per family for remote rural areas.

The 'use factor', f_u, takes into consideration the intensity of use of equipment or appliances in the different sectors, and the 'simultaneity factor', f_s, takes into consideration the probability of using that equipment or appliance simultaneously. These factors are expressed as fractions. This concept has been taken from the literature and the values have been adjusted from field observations and experience. The values of the factors are different during night and day. For example for domestic demand, the value is 0.2 for daytime and 0.9 for nighttime. This is because in most rural areas people have very low electricity consumption during the day, while during the night, and especially early evening hours their consumption is 'high'. For public lighting it is clear that during the night both factors are equal to 1 because it is expected that 100 per cent of the public lighting should be functioning, while in the daytime public lighting is off. In practice these values will be different for each site; therefore the designers should use their own judgement in choosing the values from the table or using alternatives.

Table 1.1 Use and simultaneity factors

Type of load	Nom. power (kW)	Day time			Night time		
		f_s	f_u	Maximum power	f_s	f_u	Maximum power
Domestic		0.2	0.2		0.9	0.9	
Public		0.0	0.0		1.0	1.0	
Institutional		0.8	0.8		0.5	0.4	
Industrial		0.8	0.6		0.5	0.2	
Commercial		0.9	0.8		1.0	0.9	
Special loads							
Total							

f_s = simultaneity factor, f_u = use factor

Hydrological study

The purpose of the hydrological study is to estimate the energy supply as accurately as possible, but also to make sure that the right turbine is chosen for a specific site. The latter is important in sites where there is a variable flow during the year and the turbine may have to work at partial flow; therefore careful consideration has to be given to the choice of turbine (see Chapter 9 for more details about the performance of turbines at partial flow)

It is recommended that the hydrological study is carried out by an expert in this field, i.e. a hydrologist, as the project should contain a specific document regarding the hydrology of the catchment, ending with flow diagrams clearly showing the potential over time, so that well-informed decisions can be taken.

There are various techniques for drawing up the flow diagram of a catchment. They all use flow data from several years, from which a table of absolute and relative frequency is drawn up as a function of time. Then the absolute and relative flow frequency diagrams based on time are constructed, followed by the construction of the flow duration curve whereby the probable flow occurrence can be visualized as a function of time. Figure 1.3 shows a typical flow duration curve.

The flow diagram in Figure 1.3 shows that if the hypothetical flow required for a certain design is 2.2 m³/s, the plant would work for 70 per cent of the time (70 per cent of the year, for example), whereas during the remaining 30 per cent of the time there would not be enough flow to allow the plant to operate with a full load. The flow diagram therefore allows designers to decide what to do, e.g. whether to use turbines that yield well with partial flows, use a diesel generator for all or part of this period, or manage the demand by rationing the power when the need arises. It is worth mentioning here that generally the design flow corresponds to the estimated demand (25 or 30 years), therefore for the first years of the plant's operations there may be no shortage of power, so the above mentioned estimates will be made for the future.

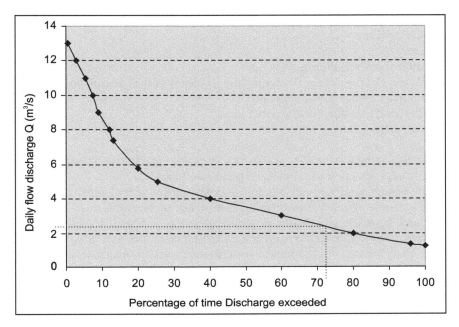

Figure 1.3 Typical flow duration curve

The estimates mentioned above are valid for small isolated schemes. However if the plant is to be connected to the grid, the analysis of the power design and therefore the design flow must be made taking the energy sales into consideration because as there is no concern about covering the demand.

Design power of a small hydroelectric plant

The design power of a small hydroelectric plant is estimated based on the demand and the evaluation of existing hydro-power resources – this aspect is not covered by this publication as there is already a considerable literature on the subject.

The two most important parameters for generating electricity are the flow (Q) and the head (H). A quick way of estimating the power of a hydroelectric plant is using the following equation:

$P = kQH$

where
P, power of the hydroelectric plant (kW)
k, factor that includes the effect of the water density, gravity and efficiency parameters

$k = \rho \cdot g \cdot \eta_p \cdot \eta_t \cdot \eta_g$

12 DESIGNING MINI AND MICRO HYDROPOWER SCHEMES

where:
ρ, water density
g, gravity
η_p, efficiency of the piping
η_t, efficiency of the turbine (including the efficiency of the transmission system)
η_g, efficiency of the power generator
H, gross head measured or estimated (m)
Q, flow (m³/s)

For these quick estimates, average k values were used, which usually differ slightly depending on the size of the plant (see Table 1.2).

The power range corresponds to the classification of hydroelectric plants. Until a decade ago, this classification comprised micro, mini, small, medium and large hydroelectric plants; now the range has increased to include pico turbines, incorporating the group of the smallest machines that can possibly be built. There are two international standards, one used by Europeans and North Americans and one used by OLADE (Latin American Energy Organisation) (see Table 1.3).

Pipes with a good internal finish and little rugosity permit a more efficient flow and therefore fewer losses. The type of material also has an influence on this; for example, in the smallest ranges, plastic pipes are used, which are highly efficient compared with steel pipes. The turbines also have a specific efficiency, depending on the finish, the manufacturing process and the model and size of

Table 1.2 Typical k values

Power range	k values
Pico plants	3.5 – 5.0
Micro plants	5.0 – 6.5
Mini plants	6.0 – 7.0
Small, larger range	7.0 – 7.5

Source: Authors

Table 1.3 Classification according to power range

Classification	Power range	
	According to OLADE	*For USA and Europe*
Pico plants	Up to 5 kW	Up to 10 kW
Micro plants	5 kW – 50 kW	10 kW – 100 kW
Mini plants	50 kW – 500 kW	100 kW – 1 MW
Small plants	500 kW – 5 MW	1 MW – 10 MW
Medium plants	5 MW – 50 MW	10 MW – 100 MW
Large plants	More than 50 MW	More than 100 MW

the machines. Similar factors influence the actual efficiency of the power generator.

Consequently, when selecting the machinery for the plant, the manufacturing quality, the background of the manufacturers and the type of turbine selected must be taken into account, in order to avoid unnecessary risks.

The hydroelectric plant

In large hydroelectric plants, the reservoir is an important component for regulating the use of water throughout the year. In small hydroelectric plants, this component is not normally used, other than in exceptional cases when the cost of this structure justifies the investment. Consequently, the following are the main components of small hydroelectric plants:

- Intake weir
- Headrace channel
- Coarse settling basin
- Conveyance channel
- Silt basin
- Forebay tank
- Penstock
- Powerhouse
- Turbine
- Generator
- Controller
- Tailrace
- Electricity grids
- Transmission grids
- Distribution grids

The electricity grids correspond to conventional electrical engineering works and are not dealt with in this text.

Similarly, it is customary to analyse a hydroelectric power plant in accordance with the following components: civil works, electromechanical equipment, electricity grids and management of the system in operation.

14 DESIGNING MINI AND MICRO HYDROPOWER SCHEMES

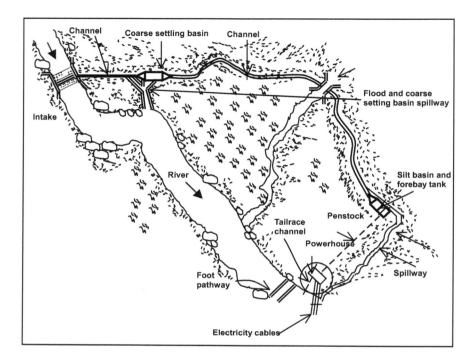

Figure 1.4 Components of micro and mini hydropower schemes

CHAPTER 2
Intake

The intake is a structure designed to capture the volume of water required for generating power and/or for other uses (irrigation, drinking water, fish farming), both during the dry season and during the rainy season. Figure 2.1 shows a plan view and cross-sections of a conventional intake used in large hydroelectric plants.

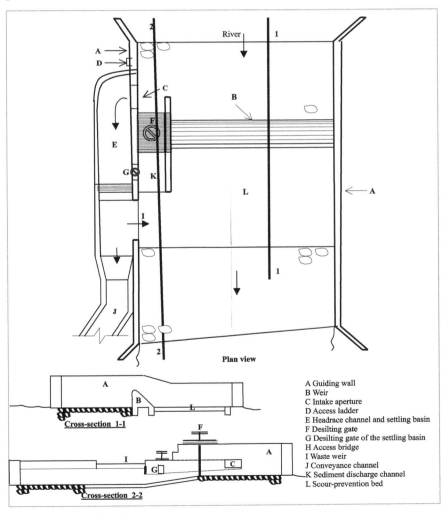

Figure 2.1 Conventional type of intake

16 DESIGNING MINI AND MICRO HYDROPOWER SCHEMES

Bearing in mind the topographic condition of the rivers, streams and/or watersheds, several types of inlet weir have been built in Peru with an appropriate, simple and safe technology, in the following locations:

- Straight section with fairly uniform topographic, hydraulic section, slope and water speed characteristics. The majority were easily accessible and easy to divert.
- Area with good geological conditions on the slopes of both river banks, compact, non-slippery soil, with no settlement or naturally protected against erosion, not prone to flooding during rainy seasons, etc.

In other cases, advantage was taken of different locations after analysing the behaviour of the watercourse:

- In a bend at the end of a straight stretch (Photograph 2.1).
- At the outlet of the pond below a waterfall.
- In the upwelling area of a spring (Photograph 2.2).

There are several types of intake. The conventional type of intake shown in Figure 2.1 has not been used in micro and mini hydroelectric plants designed and built by the authors owing to its high cost; however this type of intake is commonly used for larger hydropower schemes. For micro and mini hydro schemes requiring flows of less than 1.0 m^3/s, other types of intake of a simpler design were built, such as the intake weir with stop logs and the Tyrolean type, a description of which follows.

Photograph 2.1 Intake weir functioning with six columns and with no guiding walls

INTAKE 17

Photograph 2.2 Location of an intake weir with stop logs, by a spring. A headrace channel was not necessary, but a conveyance channel was

Intake weir with stop planks (see Annexe Example 1)

Also referred to as a fuse type or removable intake weir, because it has a removable wooden part (the stop planks) and a fixed concrete part (Figure 2.2).

The characteristics of this intake weir are:
- The height of the weir does not exceed 0.60 m. The wall of the removable weir is thin and the increasing water pressure upstream during rainy seasons and the discharge after the weir are no cause for concern.
- The intake aperture is at a higher level than the river bed, on one of the guiding walls.
- It eases operation and maintenance (fine and coarse sediments are cleared when the wooden planks are removed, taking advantage of the force of the water current).
- Locally available materials can easily be used for the construction (sand, stones, aggregate (gravel) and timber).
- The building process is simple enough for the villagers to learn.
- They are built in rivers or streams with moderate slopes, when the debris carried during rainy seasons is also moderate.
- The course of the river is fairly permanent during the rainy season.
- The cost is approximately 30 per cent cheaper than a solid concrete weir.
- The intake weirs that have been monitored so far work well both during the dry season and during the rainy season
- They are not recommended for rivers or streams with steep slopes that carry stones larger than 4" (10 cm) during the rainy season.

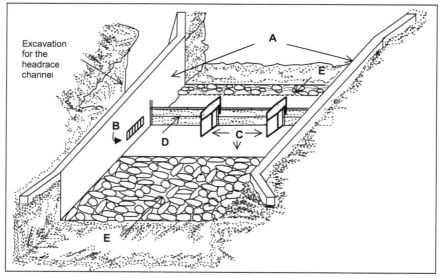

Figure 2.2 Components of an intake weir with stop planks
Note: A = guiding wall; B = intake aperture; C = fixed weir slabs and columns; D = removable weir; E = entrance (the bracing slabs go underneath the covering slabs; see details in the text).

Components

These are similar to the components of a conventional intake, although they are not all repeated, nor are they identical.

Guiding walls (Figure 2.3)

- They guide the water at a uniform speed along the section considered.
- They protect the river bank from erosion during times of flood.
- They divert the river water sideways in the case of the intake weir with stop planks, through the intake aperture.
- They maintain the stability of the slopes on both river banks.
- They contain the intake aperture and the overflow sluice.
- Their material can be simple concrete with stone or stonework contructed with sand-cement mortar.
- Their geometry can vary, although the shape that best restrains the tilting and slipping forces is preferable.
- The materials must be of good quality and built on soil with a stronger bearing capacity than the forces transmitted by their own weight and the thrust of the retained soil.
- The height will depend on the volume of the maximum flow.
- Two parallel walls are usually built, although only one is needed in some cases. When the geological conditions of the river banks in the intake area are of firm rock with no cracks, guide walls are not necessary. In that case, the intake aperture can be placed at a 45° angle from the course of the river.

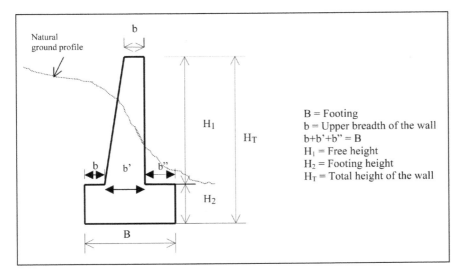

B = Footing
b = Upper breadth of the wall
b+b'+b" = B
H_1 = Free height
H_2 = Footing height
H_T = Total height of the wall

Figure 2.3 Profile of the guiding wall

20 DESIGNING MINI AND MICRO HYDROPOWER SCHEMES

Intake aperture (Figure 2.4)

- It allows the necessary quantity of river water to enter the headrace channel.
- It has a built-in steel bar safety screen to prevent stones larger than 3" (7.6 cm) or other debris from entering the headrace channel at times of high flow.
- It is built into one of the guiding walls at a higher level than the river bed and upstream from the weir.
- It is usually a rectangular section, the lower base of which should have an adequately shaped overflow profile.
- The addition of an overflow sluice is optional.

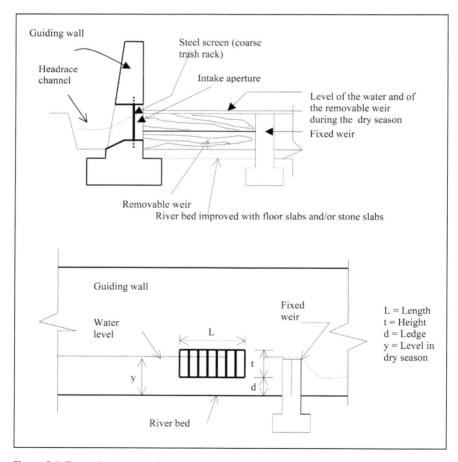

Figure 2.4 Front view and profile of the intake aperture

Weir (Figure 2.5)

- Comprises a removable part and a fixed part, located downstream from the intake aperture.
- The removable part is made of wood and the fixed part of concrete columns braced with concrete slabs and protected by a simple concrete floor slab.
- All this is located in the river bed at right angles to the guiding walls.
- The removable part is made of damp-proof wooden planks.
- The bracing slab also plays an important role, preventing the walls from sliding due to the thrust of the earth.
- The columns have a section shaped like a semi-regular hexagon to ease the flow of water and they have grooves on the sides to lodge the removable stop planks.
- The floor slab improves the river bed, providing a surface resilient to the abrasive materials carried by the river during high flows, resisting possible erosion by the water.

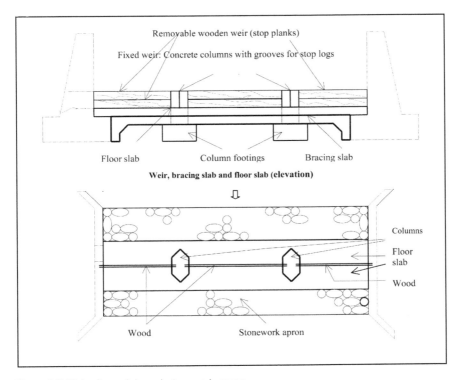

Figure 2.5 Weir, floor slab and stonework apron

22 DESIGNING MINI AND MICRO HYDROPOWER SCHEMES

Stone slabs (Figure 2.6)
- They improve the stability of the river bed and prevent erosion.
- They flatten the section of the stream in the intake area between the bracing walls next to the floor slab.
- They are located upstream and downstream from the floor slab.
- The stone slabs are settled with sand mortar and cement.
- The cross-section is shaped like an inverted U to absorb the force of the downward pressure of the water.
- They help prevent the guiding walls from sliding.

Tyrolean type intake (see Annexe Example 2)

These intakes are also called submerged weirs (Figure 2.7).
Characteristics:
- The intake aperture is located at the bottom of the river, to which end the river bed is prepared with material resistant to erosion and subsidence.
- The headrace channel is embedded in the river bed, underneath the intake aperture.

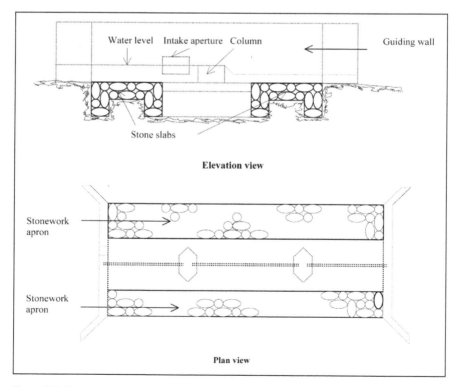

Figure 2.6 Stonework apron with mortar of cement and fine sand

- It may or may not have a low weir. It does not have a weir when the intake aperture is as wide as the river and the water pressure is steady all year round, varying only during the rainy season. It does have a weir when it covers only part of the width of the river and the water pressure varies to a minimum pressure during the dry season.
- The weir is very low, in many cases no more than 0.20 m high, therefore the increase in the upstream pressure during the rainy season is normal.
- It must be located in a straight section and the difference between its level and that of the settling basin located a few metres from the intake is very important. Likewise, the level of the settling basin with respect to the level of the river during high flows is important for evacuating sediments.
- The slopes of the rivers or streams must be greater than 4 per cent and their course must be steady with no change of direction.
- It provides a free passage for the debris dragged by the river during the rainy season. During the dry season the weir must be cleaned, particularly in autumn when the water carries floating leaves.

Components

The following are the characteristics of Tyrolean intake components:

Guiding walls: The same as the intake weir with stop planks
Intake aperture (Figure 2.8)

- This is an area located at the same level as the river bed (bottom) with a steel bar screen (trash rack) embedded in the top of the collecting channel.
- The steel bars can be selected in accordance with the efficiency and type of section to be used.

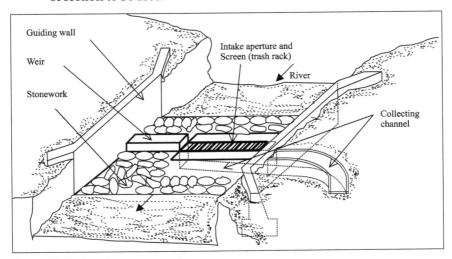

Figure 2.7 Tyrolean type intake weir

24 DESIGNING MINI AND MICRO HYDROPOWER SCHEMES

- The steel bars are placed in the direction of the water current.
- The inclination of the screen will depend on the slope of the river in the selected section.
- It can be as long as the entire width of the river or only cover part of it, depending on the design.
- It is easy to clean.

Collecting channel (Figure 2.9)

- It is located underneath the intake aperture and is totally embedded in the river bed.
- The water it will carry is the flow obtained through the screen.
- The selected slope must be such that it will not allow any sand sediments or gravel to go through the screen, particularly during the rainy season.
- The construction material should preferably be concrete, with casing and a polished finish.
- It connects directly with the outside headrace channel and with a coarse settling basin, to evacuate thick debris towards the river.

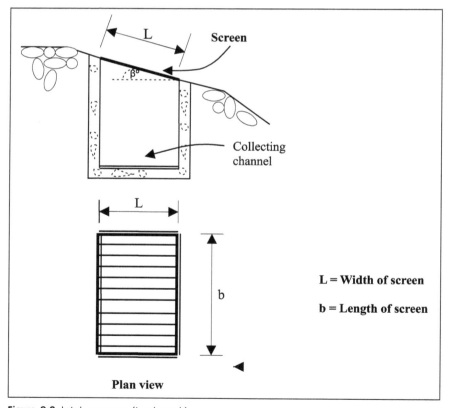

Figure 2.8 Intake screen (trash rack)

Weir (Figure 2.9)

- The weir is designed when the intake screen (trash rack) does not cover the entire width of the river.
- During the dry season, it forces the water in the intake to flow towards the intake screen.
- Its height above the intake screen is usually less than 0.20 m and it is made of abrasion-resistant concrete.
- It is a concrete dyke embedded in the river bed alongside the headrace channel.
- When this type of intake has a fixed weir, it must be made of simple resilient concrete and of a low height.

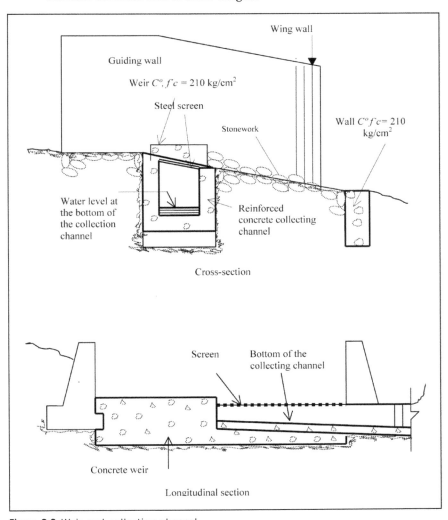

Figure 2.9 Weir and collecting channel

26 DESIGNING MINI AND MICRO HYDROPOWER SCHEMES

Stonework apron

- This is placed upstream and downstream from the collecting channel and/or weir. It has the same characteristics as the intake weir with stop planks.
- At the end of the stonework apron downstream, an embedded concrete wall is required across the centre line of the river, to guarantee its durability and prevent rapid underscouring of the river bed.

Complementary components

The headrace channel and the settling basin (Figure 2.10) contribute to the normal operation of the intake, both during the dry season and during the rainy season. Their functions are described in the following chapters.

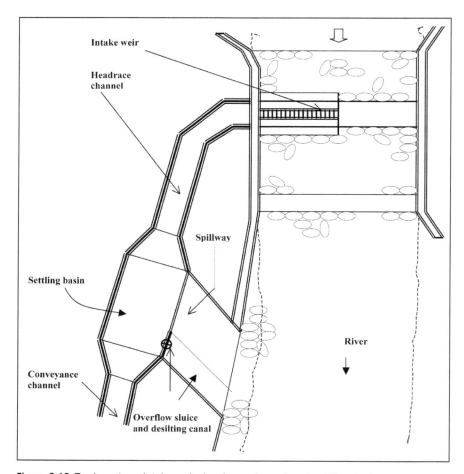

Figure 2.10 Tyrolean type intake weir, headrace channel and settling basin

INTAKE   27

Design of the main components

Guiding walls

They can be made of stone masonry mixed with cement mortar and sand, coarse concrete, etc., designed to support their own weight and lateral loads caused by the thrust of earth. They need no reinforcement and should withstand the forces on them by their own weight (Figure 2.11).

Forces and pressures acting on the wall (Figure 2.12)

Downward forces:
- P_m = weight of the wall
- P_s = weight of the soil on the wall at the back of the core wall

Upward forces:
- N = pressure of the soil at the base of the wall

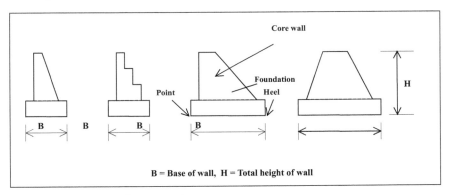

Figure 2.11 Examples of cross-sections

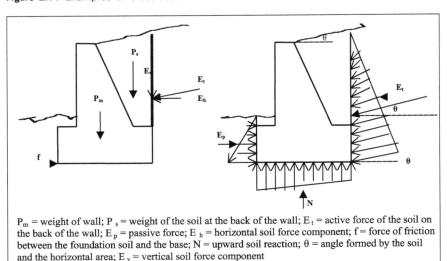

P_m = weight of wall; P_s = weight of the soil at the back of the wall; E_t = active force of the soil on the back of the wall; E_p = passive force; E_h = horizontal soil force component; f = force of friction between the foundation soil and the base; N = upward soil reaction; θ = angle formed by the soil and the horizontal area; E_v = vertical soil force component

Figure 2.12 Diagrams of the forces and pressures acting on the wall

28 DESIGNING MINI AND MICRO HYDROPOWER SCHEMES

Lateral forces:
- E_t = active force of the soil at the back of the sloping core wall. It has horizontal and vertical components if $\theta \neq 0°$
- E_p = passive force of the soil on the front of the wall footing. This is not usually considered because it is changed for bracing slabs, floor slabs and/or stonework aprons.
- f = force of friction between the soil and the base of the wall.

Wall design

The type of construction material to be used must be determined. When coarse concrete is used, no reinforcement is needed and the wall can be as high as 3 m with a variable length.

The design begins with (tentative) dimensions of the selected cross-section, for a 1 m long wall, bearing in mind the following (referential) criteria:

a) $0.5H \leq B \leq 0.7H$
b) $b \geq 0.30$ m (to make it easier to lay the concrete and the large stones)
c) $H = H_1 + H_2$
   H_1 = height of the core wall
      = Thickness of the floor slab + H WEIR + $Y_{max\ flow}$ + free border
   H_2 = footing height, depending on the quality of the foundation soil
   $0.35 \leq H_2 \leq 0.80$

The height or pressure during high flows is determined by the hydrological study. It can also be obtained by taking field data in accordance with the marks left by the water during heavy rainfall, in normal years and in years when exceptionally heavy rainfall occurs.

d) Identify the soil characteristics: types, cohesion, friction coefficients, internal friction angle, specific weight, etc. The angle formed by the

Table 2.1 Specific soil weights and friction angles

Type of soil	Specific weight γ_s (t/m³)	Friction angle, ø (°)
Dry embankment soil	1.40	37
Damp embankment soil	1.60	45
Wet embankment soil	1.80	30
Dry sand	1.60	33
Damp sand	1.80	40
Wet sand	2.00	25
Dry diluvia slime	1.50	43
Damp diluvia slime	1.90	20
Dry clay	1.60	45
Damp clay	2.00	22
Dry grit	1.83	37
Damp grit	1.86	25
Angular gravel	1.80	45
Boulder gravel	1.80	30

Source: Terán (1998)

Table 2.2 Types, admissible forces and friction coefficients of foundation ground

Foundation ground		Admissible force, σ_t (kg/cm²)	Friction coefficient, μ
Rocky	Uniform hard rock with few cracks	10.00	0.70
	Hard rock with many cracks	6.00	0.70
	Soft rock	3.00	0.70
Gravel stratum	Dense	6.00	0.60
	Not dense	3.00	0.60
Sandy ground	Dense	3.00	0.60
	Not dense	2.00	0.50
Cohesive ground	Very hard	2.00	0.50
	Hard	1.00	0.45
	Medium	0.5	0.45

Source: ACI, 1998

slope and the horizontal area will be measured in the field. Tables 2.1 and 2.2 show the values obtained in laboratories to help design the wall. When the soil is different from those soil types in these tables, it is advisable to carry out tests in the soil mechanics laboratory.

e) Wall stability conditions
   1. Bending stability:

$$\frac{\Sigma M_e}{\Sigma M_v} \geq 2$$

ΣM_e = sum of stabilizing force moments
ΣM_v = sum of bending force moments

   2. Sliding stability:

$$\frac{f}{E_t} \geq 2$$

$f = \mu N$

$f =$ force of friction between the foundation soil and the concrete base of the wall
$\mu =$ friction coefficient
$N =$ total reaction of the soil at the base of the wall

$$E_t = \frac{1}{2} \times K_a \times \gamma \times H^2$$

$E_t =$ Active thrust of the soil that causes sliding
$K_a =$ Soil thrust coefficient

$$K_a = \cos\theta \times \frac{\cos\theta - \sqrt{\cos^2\theta - \cos^2\phi}}{\cos\theta + \sqrt{\cos^2\theta - \cos^2\phi}}$$

$\theta =$ angle formed by the inclination of the soil with the horizontal bank; in some cases this value is zero
$\phi =$ internal soil friction angle (laboratory tests)

γ = specific soil weight, either in (N/m³ or in kg/m³)
H = total height of the wall (m).
3. Stability of the foundation soil:
This consists of verifying:

$\sigma_{\text{admissible soil factor}} > \sigma_{\max} > \sigma_{\min}$

$$\sigma_{\min} = \frac{N}{B}\left[1 - \frac{6 \times e}{B}\right]$$

$$\sigma_{\max} = \frac{N}{B}\left[1 + \frac{6 \times e}{B}\right]$$

$\sigma_{\min}$ = minimum force of the wall and part of the footing soil on the foundation soil (kg/cm²)
$\sigma_{\max}$ = maximum force of the wall and the soil together on the foundation soil (kg/cm²)
B = base of the wall (m)
e = eccentricity (m) $e = \dfrac{\Sigma Mo}{N}$
ΣMo = sum of the momentum of all the forces involved in the centre of the base: stabilizing and bending moments
N = total reaction of the soil on the base of the wall

Tyrolean type intake screen

The following formulas, value tables and calculation procedures were used to design this type of intake (see more information in Lauterjung and Schmidt, 1989); see sketch of this type of intake in Figure 2.13.

Formula for calculating length b and width L of the screen

$$Q = \frac{2}{3} \cdot c \cdot \mu \cdot b \cdot L \cdot \sqrt{2gh}$$

Where:
Q = Flow in m³/s passing through the screen
$c = 0.6\, \dfrac{a}{d} \cdot (\cos\beta)^{3/2}$
a = distance between bars
d = distance between bar centres
$\beta°$ = angle formed by the screen and the horizontal
Recommendation: $5° < \beta < 35°$

 Rectangular section bars

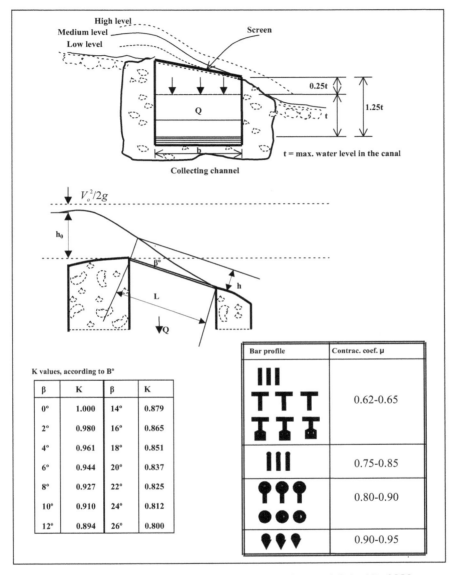

Figure 2.13 Planning of intake structures. *Source:* Lauterjung and Schmidt, 1989

μ = contraction coefficient, according to the steel profile of the screen
b = length of the screen across the river
L = width of the screen

$$h = \frac{2}{3} \cdot k \cdot h_0$$

k = h correction coefficient, according to angle β (see Figure 2.13)
h_0 = river level before reaching the screen

Intake construction procedures

It is not the intention to provide a recipe, but to report the steps used to build the intake.

Verification of drawings

This step consists of reviewing the plans of the intake and identifying the consistency of the dimensions, technical specifications, design details, as well as of the components that immediately follow the intake.

For example, to ensure that the topographic plan contains level curves, elevations, plan views and crosscuts; that level details and dimensions are consistent; that no technical specifications are missing and that there are no contradictions between the descriptive report, the design criteria, the hydrological study, etc.

Survey of the intake site

Accessibility

In most cases the only access to the intake is a simple path; it may or may not be necessary to clear rubble or carry out complementary works such as provisional or definitive pedestrian or carriage bridges to facilitate the transport of materials and equipment and provide access for construction workers.

Location of the intake and its surroundings

Verify:
- Appropriate location
- River slope
- Type of soil on the river bed
- Main elevation of the river bed, elevation of the bottom of the headrace channel, settling basin, etc.
- Compare the bearing capacity of the foundation ground specified in the plan with the type of soil identified (Table 2.2).
- Stability of both river banks: natural protection of the slopes against erosion (natural vegetation, angle of slope, presence of run-off and/or formation of ditches, etc.).
- Select an appropriately safe area as a storeroom for materials and tools, a dressing room and temporary sanitary fixtures. In some cases, renting a nearby home can be a solution.
- Select an appropriate area for receiving aggregates and preparing and mixing measured quantities.

Layout and stake out

With the help of the in-plant plan of the intake, carry out the following tasks.
- Clear the area of vegetation and remove stones from the river course.
- Draw the central line of the intake in the river bed, using adequately placed large stakes.
- Lay out the guiding walls and the centre of the weir, with survey poles, fixed stakes or painted signs on fixed river components.
- Locate elevations or key levels: finished floor of the fixed weir, elevation of the bottom of the headrace channel, etc., with the help of topographic equipment.
- Stake out the corresponding layout and levels.

See Figure 2.14

Construction of the guiding walls

Before building one of the walls, temporarily channel the river to half its original course, using sandbags or wood (Figure 2.15). Start with the guiding wall that will contain the intake aperture, then follow the steps below.

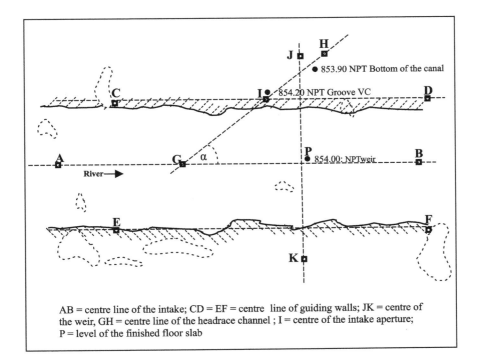

Figure 2.14 Layout of central lines and levels in the intake

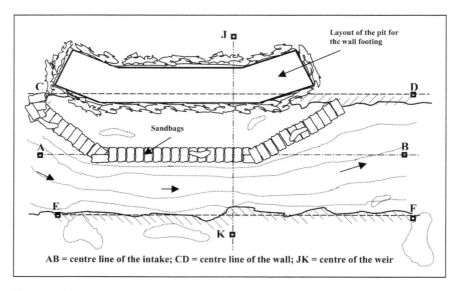

Figure 2.15 Temporary channelling of the river and excavation of the footing pit

Construction of the footing

- Lay out the footing, taking centre lines CD and AB as a reference.
- Dig the footing pit to the specified depth.
- Clear the land of any debris.
- Establish the elevation of the foundation and the footing (footing height).
- Formwork, if necessary.
- Empty the water in the footing pit with drains or pumps, if necessary.
- Set aside an area near the footing pit to prepare the mix.
- Constantly supervise the dosage and mix of the materials and the adequate transportation of the concrete in tins or buggies, trying to prevent the concrete from sweating.
- Lay the concrete, shifting large stones to make sure they are always lined with concrete until the specified elevation or height.
- It is advisable to leave stones sticking out by one-third, like teeth, on the upper surface of the footing, separated and aligned so that they can then be joined to the body of the wall. See Figure 2.16

Construction of the body of the wall

This comprises the layout, formwork and laying of the concrete.
- On the top surface of the footing, plot the width of the base, taking centre line CD as a reference.
- Prepare the forms for the entire wall, in accordance with the measurements, including the wing walls.

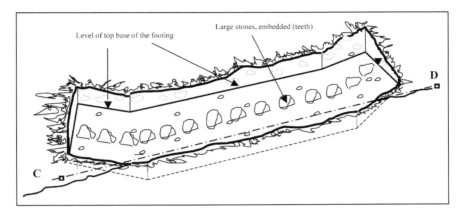

Figure 2.16 Excavation and casting of the wall footing

- Carry out the formwork, supervising its rigidity, verticality, height, alignment, gradient along the centre line CD and the design inclination of the back of the wall. Also the formwork, bracing, etc. of the whole wall, in accordance with the wall forming technology and the project's technical specifications.
- Draw guidelines inside the wall form so as to design the form for the intake aperture.
- Prepare a sufficiently rigid form for the intake aperture and fit in properly spaced steel bars.
- Place the form of the intake aperture at the level of the ledge and, in accordance with the guidelines drawn, ensure that it is plumbed and levelled at the right time during the laying of the concrete. It must be fixed to prevent shifting and rigid to avoid any deformity caused by the pressure of the fresh concrete.
- Wet the inside of the form with petrol to preserve the wood and ease the formwork.
- In the upper part at the back, leave a big enough access so that at least one person can enter to accommodate the concrete on the base.
- Pour on the concrete, under supervision, applying concrete technology standards.
- Strip the forms after at least 48 hours if the temperature is more than 12°C and then cure the concrete (Figures 2.17 and 2.18).

The other guiding wall is built in a similar way, staking out the corresponding layout and always taking centre line AB as a reference and temporarily channelling the course of the river so that it flows alongside the already built wall.

36 DESIGNING MINI AND MICRO HYDROPOWER SCHEMES

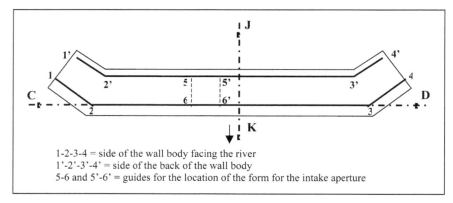

Figure 2.17 Layout of the footing to form the wall body and the intake aperture

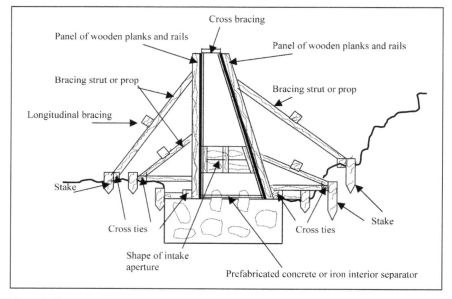

Figure 2.18 Forming of guiding wall and intake aperture

Weir

Before building the second guiding wall, it is convenient to use the first channel to build half of the weir, following the steps below:

Construction of the column footings

- Lay out and stake out the dimensions considered in the plan and on the land.
- Excavate and remove the dug out material.

- Establish the foundation level.
- Fit in the corresponding lengthwise, crosswise and vertical steel bars of the footing and the column.
- Align, plumb and attach all the steel before pouring the concrete.
- Establish the height of the footing.
- Pour the concrete to obtain the specified height.

All these steps must be taken in accordance with the technical specifications and details indicated in the plan (diameter of the steel, fluency limit, steel casing, distribution of abutments, concrete quality and dosage, admissible force on the foundation soil, etc.). See Figures 2.19 and 2.20.

Construction of the bracing slab (Figure 2.21)

- Plot and stake out the measurements.
- Excavate to the specified depth, shaping the designed cross-section and removing the dug out material.
- Fit in the crosswise and lengthwise steel bars, in accordance with the spacing established in the detailed plan.
- Join the lengthwise steel bars in accordance with the area and the length established in the plan.
- Level and secure the steel for the corresponding covering.
- Establish the level of the top of the bracing slab.
- Pour the concrete, in accordance with the technical specifications.

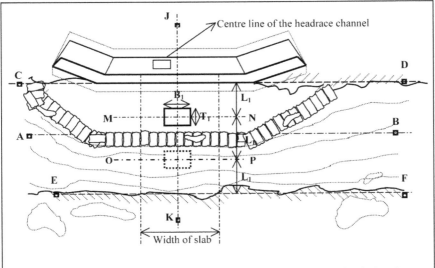

AB = centre line of intake structure; CD = centre line of wall; JK = centre of weir; $L_1 = L_2 = L_3$ = distance between the centre lines of columns; MN and OP = centre lines across and at right angles to the centre of the weir; B_1 and T_1 = dimensions of column footings

Figure 2.19 Layout of column footings for excavation purposes

38 DESIGNING MINI AND MICRO HYDROPOWER SCHEMES

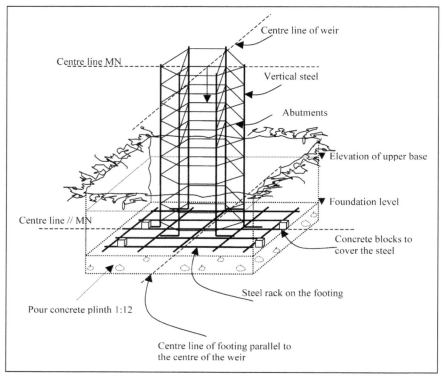

Figure 2.20 Plinth, reinforced steel in footing and column, determination of foundation levels and laying of concrete

Finish the construction of the columns and build the cover slab (Figure 2.21)

- Repeat the lay out and stake out, always with the same accurate measurements and levels.
- Prepare the wood, giving it the geometric shape of the column section, putting special emphasis on the column grooves.
- Clean the steel on the column, removing all traces of concrete.
- Form the column or columns, securing their verticality, gravity and alignment with the central lines of the weir and between columns. Also verify the verticality and correct alignment of the wooden planks that will shape the grooves on the notches.
- Pour the concrete on the column, making sure there are no holes.
- Strip the form after at least 24 hours, trying not to alter the alignment.
- Cure the concrete three or four times a day continuously for at least eight days.
- Form the borders so as to form the cover slab.
- Put in place the levels on the upper side of the cover slab.
- Pour the slab concrete up to the specified height.
- Cure the slab.

Add the finishing touches to the columns and the intake aperture

- Complete to the final design measurements.

Construction of the stonework apron

- Plot and stake out the measurements.
- Excavate to the specified depth to form the stone clusters at both ends and remove the dug out material.
- Establish the upper surface levels.
- Form the stone clusters until reaching the total height (upper surface), as though they were foundations of cement, aggregate and stone 1:8 + 25% large stones.
- Lay a concrete cement floor before bedding the stone.
- Bed the recommended stone, forming a uniform surface. Leave spaces of 3 to 5 cm between each stone.
- Set the stone with mortar of cement and fine sand in the proportion 1:3, making sure that all the joints are completely filled in. Use appropriate tools.
- Cure the set stone work.

The above corresponds to one part of the intake construction. The other part can be built in a similar way. If it is easy to channel the river outside the working area of the intake, then all the steps mentioned above should be followed.

Construction of the removable weir

- Select the timber in accordance with the specified measurements, thickness and quality.
- Try installing it in the fixed weir, making sure there are no imperfections in either the column groove or the wood which would make it difficult to install and dismantle.
- Grease it with suet to protect it from dampness and make it more durable.

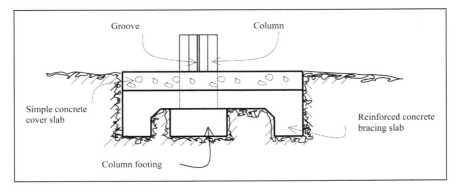

Figure 2.21 Fixed weir (footing, column, bracing slab and cover slab)

Technical specifications

These are specific characteristics that guarantee the quality of the design, the building process and the estimated materials for each of the civil works components. They can usually be found in the National Construction Regulations of each country and are detailed in the descriptive report and in each of the project's plans. As examples see Tables 2.3 to 2.11, taken from the Peruvian National Standards Institution.

Density of materials

Table 2.3 Density of materials in kg/m^3 (National Construction Regulations)

	Density in kg/m^3
Metals	
Construction steel	7,850
Binders	
Cement	1,450
Aggregates	
Soil	1,600
Gravel and dry sand	1,600
Concrete (sand + stone)	1,880
Liquids	
Water	1,000
Oil	870
Masonry	
Limestone	2,400
Brickwork	
Solid bricks	1,800
Hollow bricks	1,350
Adobe	1,600
Coating with mortar and	
Cement	2,000
Gypsum	1,000
Mud	1,600
Concrete	
Simple or coarse	2,200
Reinforced concrete	2,400
Timber	
Dry coniferous	450
Damp coniferous	650
Hard (dry)	700
Hard (damp)	1,000

	Unit mass in kg/m^2
Roofing	
Non-industrial baked clay tiles	160
Zinc sheeting	12

Quality of the concrete determined by its f'_c resilience (to compression at 28 days in kg/cm^2)

The following quantities were used for guiding walls, footings, columns, slabs, stone work, foundations, continuous footing, channels, casings, mortars, floors, pavements, anchors, foundations for the base of the turbine and generator, etc., with good results in terms of durability, anti-abrasion and workability.

Table 2.4 Contributing coefficients for 'in volume' mixes, with 5% waste per m^3 of concrete

Cement : Aggregate (Type I Portland cement) 1 sack of cement (C) = 42.5 kg, loose aggregate (Ag) in m^3				
1:6	1:8	1:10	1:12	1:14
C : Ag(m^3)	C : Ag(m^3)	C : Ag(m^3)	C : Ag(m^3)	C : Ag(m^3)
6.30:1.25	4.70:1.25	3.70:1.25	3.00:1.25	2.65:1.25

Table 2.5 Contributing coefficients for 'in volume' mixes, with 5% waste per m^3 of concrete

Cement : Aggregate : Stone
(Type I Portland cement)
1 sack of cement (C) = 42.5 kg, loose aggregate in m^3(Ag), loose stones in m^3(S)

Type of mix	Percentage of stone in 1 m^3 of concrete								
	25%			30%			40%		
	C (sack)	Ag(m^3)	S (m^3)	C (sack)	Ag(m^3)	S (m^3)	C (sack)	Ag(m^3)	S (m^3)
1:6	4.72	0.94	0.31	4.41	0.88	0.38	3.78	0.75	0.50
1:8	3.52	0.94	0.31	3.29	0.88	0.38	2.82	0.75	0.50
1:10	2.78	0.94	0.31	2.59	0.88	0.38	2.22	0.75	0.50
1:12	2.25	0.94	0.31	2.10	0.88	0.38	1.8	0.75	0.50
1:14	2.00	0.94	0.31	1.59	0.88	0.38	1.59	0.75	0.50

Stone porosity, 25% average and 4"<effective diameter < 8"

Table 2.6 Contributing coefficients for 'in volume' mixes, with 5% waste per m^3 of concrete

Cement : Sand : Stone
(Type I Portland cement)
1 sack of cement = 42.5 kg, thick loose sand in m^3, split and crushed loose stone in m^3

f'_c	Cement (sack)	Sand (m^3)	Stone (m^3)	Water (gallon/sack)	Proportions Cement: Sand: Stone
140	7.88	0.546	0.827	7.5	1:2:4
175	9.00	0.472	0.717	6.75	1:2:3
210	10.00	0.478	0.552	6.00	1:2:2

Table 2.7 Contributing coefficients for sand-cement mortar, in volume per m³ (including 5% waste)

Proportions	Thick sand				Fine sand				Average	
	25% Voids		30% Casting		35% Casting		40% Casting			
	Cem. (sack)	Sand (m³)	Cem. (sack)	Sand (m³)	Cem. (sack)	Sand (m³)	Cem. (sack)	Sand (m³)	Cem. (sack)	Sand (m³)
1:1	21.20	0.68	21.80	0.56	22.50	0.53	23.20	0.55	22.18	0.58
1:2	14.83	0.79	15.46	0.79	16.13	0.77	16.90	0.80	15.83	0.79
1:3	11.40	0.90	12.00	0.92	12.57	0.89	13.20	0.94	12.29	0.91
1:4	9.26	0.99	9.75	1.00	10.30	0.97	10.90	1.04	10.05	1.00
1:5	7.80	1.02	8.24	1.05	8.73	1.03	9.30	1.09	8.52	1.05
1:6	6.74	1.07	7.12	1.10	7.57	1.08	8.10	1.14	7.38	1.10
1:7	5.93	1.10	6.27	1.13	6.62	1.11	6.60	1.19	6.36	1.13
1:8	5.29	1.13	5.61	1.14	5.99	1.12	6.40	1.22	5.82	1.15

Table 2.8 Contributing coefficients for primary plastering with cement and sand mortar, per m² (including 5% waste)

Mortar	Thickness 0.5 cm		Thickness 1.00 cm		Thickness 1.2 cm		Thickness 1.5 cm	
	Cem. (sack)	Sand (m³)	Cem. (sack)	Sand (m³)	Cem. (sack)	Sand (m³)	Cem. (sack)	Sand (m³)
1:1	0.111	0.003	0.222	0.006	0.266	0.007	0.333	0.009
1:2	0.080	0.004	0.160	0.008	0.192	0.010	0.240	0.012
1:3	0.062	0.005	0.123	0.009	0.148	0.011	0.185	0.014
1:4	0.051	0.005	0.101	0.009	0.121	0.011	0.152	0.014
1:5	0.043	0.005	0.085	0.009	0.102	0.013	0.128	0.016
1:6	0.037	0.006	0.074	0.011	0.089	0.013	0.111	0.017

Table 2.9 Contributing coefficients for whitewashing with pure gypsum per m² (including 5% waste)

Thickness	0.5 cm	0.7 cm	1.0 cm	1.2 cm	1.5 cm	1.8 cm
Gypsum (kg)	5	7	10	12	15	18

Table 2.10 Contributing coefficients for concrete floors, pavements or floor plates per m² (including 5% waste)

Thickness	2" (5.1 cm)		3" (7.6 cm)		4" (10.2 cm)	
Mix	Cement (sack)	Aggregate (m³)	Cement (sack)	Aggregate (m³)	Cement (sack)	Aggregate (m³)
1:6	0.32	0.064	0.48	0.095	0.64	0.127
1:8	0.24	0.064	0.36	0.095	0.48	0.127
1:10	0.19	0.064	0.28	0.095	0.38	0.127
1:12	0.15	0.064	0.23	0.095	0.30	0.127
1:14	0.13	0.064	0.20	0.095	0.27	0.127

Table 2.11 Resistant concrete floors of cement, sand and stone, per m², (including 5% waste)

Thickness	4" (10.2 cm)			5" (12.7 cm)			6" (15.2 cm)		
f'_c (kg/cm²)	Cem. (sack)	Sand (m³)	Stone (m³)	Cem. (sack)	Sand (m³)	Stone (m³)	Cem. (sack)	Sand (m³)	Stone (m³)
140	0.80	0.048	0.096	1.00	0.060	0.120	1.20	0.072	0.14
175	0.96	0.051	0.084	1.20	0.072	0.105	1.44	0.086	0.13
210	1.02	0.055	0.083	1.28	0.089	0.104	1.54	0.083	0.13

CHAPTER 3
Headrace channel

The water obtained by the intake is conveyed through the headrace channel. Since the flow is greater during the rainy season than during dry season, it carries a surplus flow during that period. The channel begins behind the intake aperture or trash rack and plays an important role, providing stability by evacuating excess water and material (gravel, sand) that flows through the screen.

The headrace channel comprises the channel itself and the spillway as a complementary component.

Channel

The channel's characteristics are determined by the geometrical shape assigned to it, which must respond satisfactorily to the hydraulic characteristics.

The geometric components (foundation, total height, free border and breadth), the hydraulic characteristics (slope, brace, hydraulic radius, wet area, wet perimeter and roughness) and other important characteristics are similar to those of the conveyance channel which are detailed in Chapter 5.

Design of the headrace channel

The channel design is determined once the dimensions of the trash rack in the intake have been calculated, both for low water and high flow stages. See Example 1 of Annexe I, in which a 0.9 m breadth was obtained for a 6 per thousand slope. For this case, the hydraulic section is described in Figure 3.1.

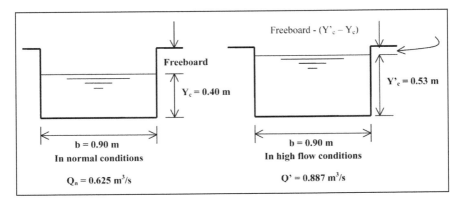

Figure 3.1 Hydraulic section of the headrace channel

The free border is determined by calculating the dimensions of the spillway or side overflow channel.

Building procedure

The construction of the headrace channel begins once the trash rack is completed.

In some cases, preliminary work is required due to the topography and geology, such as the use of explosives if the area is rocky, drilling the rock to obtain a headrace channel, constructing a special structure suitable for the natural conditions of the area, or building an overhanging headrace channel, for example.

The following steps refer to areas where no explosives or special works are required, as the ground is easy to excavate with reliable corner angles. This has been summarized in three important steps, each comprising specific tasks.

Construction of the platform

- Clear the area of any shrubs and/or weeds using hand tools.
- Determine the total width of platform AB, in accordance with the details of the channel. Figure 3.2 (d): AB = x_1 + 2e + b + x_2, where:
  x_1 = external crown
  e = breadth of the channel slope
  b = width of the channel
  x_2 = internal crown
- Draw the outside border of the platform (a polygon) with topographic equipment, starting from the trash rack, with the same slope as the channel and putting stakes 1, 2, 3 in place (Figure 3.2 (a)).
- With lime or gypsum, draw the inside border of the platform, based on the AB width of the platform and using the previous step as a reference (Figure 3.2 (b)).
- Using hand tools, cut out the required slope along the plotted inside border, checking the depth with the topographic level and putting stakes 1, 2, 3, ... in place (Figure 3.2 (c)).
- Remove the excavated material to the selected areas.
- Lay out the platform levels every 10 m in straight sections and every 5 m in curves.

Construction of the channel (Figure 3.3)

- On the built platform, plot the centre line of the channel and stake out the polygon obtained.
- Draw the curve at the junctions with the polygon, in accordance with the indicated curvature radius.

HEADRACE CHANNEL 47

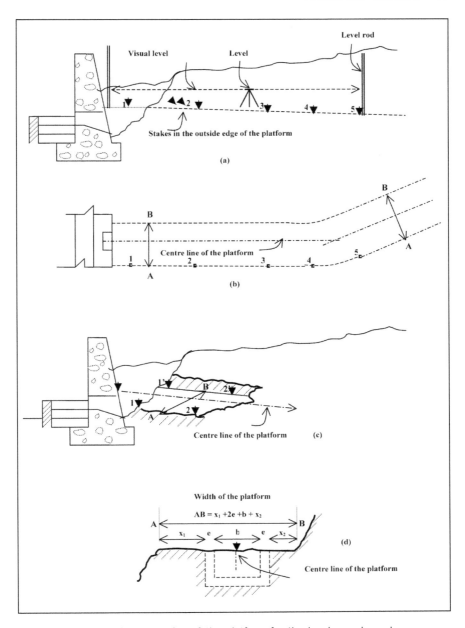

Figure 3.2 Layout and construction of the platform for the headrace channel

- Draw the channel borders, bearing in mind the total width and thickness of the embankments = b + 2e.
- Dig the channel to the required depth, in accordance with the type of section and slope design, including the breadth of the bottom and the embankments.
- As the excavation progresses, put the guides in place at the height level required by the slope and the bottom of the channel, every 10m for straight sections and as dependent on the curvature radius for curved sections.
- Shape the excavation and/or remove large stones from the dug out channel.

Lining and waterproofing the channel

- Stake out the central line of the channel and mark it by tying rope around the bollards.
- Put guide stakes in place at the bottom of the channel to give it the designed slope and floor thickness.
- Apply the channel forming technology in accordance with the coating method selected, securing the formation of expansion joints.
- Apply the concrete technology to coat the channel in accordance with the forming method selected (trussed beam method or the conventional method, etc.).
- Strip the forms and add the finishing touches in accordance with the technical specifications.
- Cure for a minimum of eight days.
- Fill in the expansion joints with asphalt and fine sand, in the correct proportions and following the technical specifications.

Characteristics and technical specifications

- Quality of concrete: $f'_c = 175$ kg/cm^2
- Thickness: 0.18 m (minimum)
- Slope S: 6 per thousand

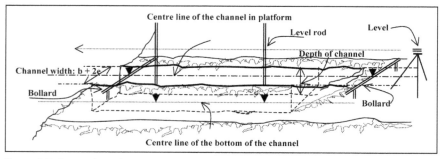

Figure 3.3 Excavation of the channel

- Polished finish   3:1 (sand:cement)
- Length   40 m
- Spillway   Left bank, before the trash rack
- Geometric characteristics in accordance with Figure 3.1

Spillway (see Annexe Example 3)

The spillway is a U-shaped section for the evacuation by gravity of the surplus water from the channel to adjacent areas (Figure 3.4). In the majority of cases, the spillways need other accessories, such as flood gates and overflow channels, to prevent erosion or any other type of damage on the land crossed by the channel. Micro and mini hydropower schemes generally have more than two spillways located along the channel from the intake to the forebay tank.

The spillway is usually located alongside one of the walls of the channel, the settling basin or the forebay tank. For example:
- Before the end of the headrace channel, to evacuate the surplus flow during flood conditions.
- In some sections of the conveyance channel, where it is anticipated that a landslide may cause an uncontrollable overflow.
- Before the settling basin so that the water can be diverted to the spillway for maintenance and cleaning purposes.

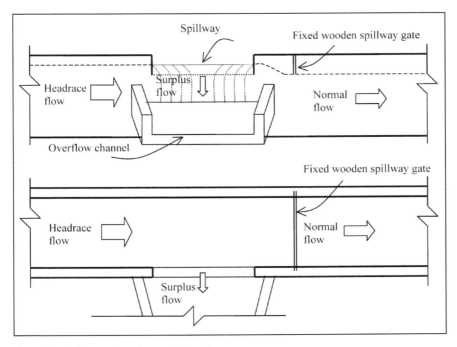

Figure 3.4 Spillway plan view and elevation

50 DESIGNING MINI AND MICRO HYDROPOWER SCHEMES

- In the forebay tank to evacuate all the water from the turbine when the operator shuts off the valve in an emergency, or if the water in the nozzle is suddenly obstructed by foreign matter entering the penstock (e.g. in the Pelton turbine which only has one nozzle).

The spillway elements (see Figure 3.5) are basically:
- width (L);
- head (h);
- free border (t').

Spillway design

In the case of the headrace channel, the following information is required:
- depth during flood conditions (Y'_c);
- headrace flow (Q');
- design flow (Q_n);
- depth during typical flow (Y_c);
- surplus flow (Q).

Subsequently, apply the standard spillway formula:

$$Q = C \times L \times h^{1.5}$$

where: C = discharge coefficient = 1.6

Construction procedure

- Identify the place in the channel where the spillway should be built.
- Plot the dimensions of the spillway on the formwork of the channel slope: L and height. These measurements must include the final coating.
- When finishing, give it the crest shape shown in the plan.
- Supervise the laying of the concrete in the correct proportions.
- Strip the form and add the specified finishing touches.
- Carry out tests to make sure it works (during high flow stages).

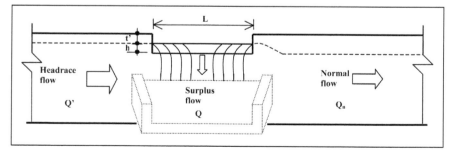

Figure 3.5 Spillway in elevation

Technical specifications

The technical specifications of the spillway are the same as those of the headrace channel, as it is a diversion in one section of one of the embankments.

The crest shape of the spillway is specified and the section can be rectangular or trapezoidal.

In most cases, the spillways are complemented with an overflow channel that receives the surplus water evacuated by the spillway and conveys it adequately without causing any damage to adjacent land, plantations, etc. The overflow channel is described in Chapter 6 as a complementary component of the silt basin.

Photograph 3.1 shows the spillway alongside the headrace channel of the Santo Tomás de Cutervo MHS in Cajamarca, Peru.

Photograph 3.1 Headrace channel

CHAPTER 4
Coarse settling basin

The construction of this structure is the same as that of the headrace channel and similar to that of the silt basin. Its purpose is to eliminate granular material larger than the sand or gravel that passes through the headrace channel during flood conditions. The construction material is concrete, resistant to the abrasive effect of floating debris. During the wet season, the sluice gate remains slightly open so that the gravel does not settle but is constantly expelled instead of passing through to the conveyance channel.

The settling basin is located:
- after the headrace channel
- preferably on a straight stretch
- The level at the bottom of the crest weir should be higher than the high flow level of the river.
- It should not be exposed to an area prone to cave-ins or landslides.
- The bearing capacity of the foundation soil should be identified. There should be no settlement.

The elements are the same as those of the silt basin, which is described in more detail in Chapter 6. However, the weir is as wide as the channel, whereas the silt basin is wider (see Figure 4.1).

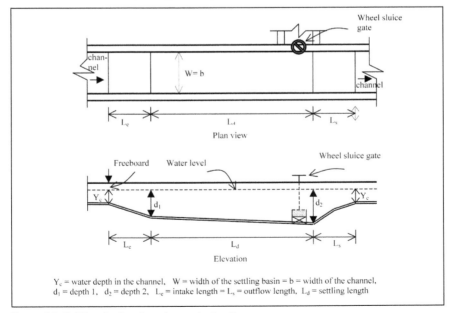

Figure 4.1 Settling basin plan view and elevation

Design (see Annexe Example 4)

The design of the settling basin consists of determining the inner dimensions of the basin. The main parameters are calculated using the equations below; however some technical data are obtained in the field. Once the inner dimensions have been estimated, a minimum breadth and the recommended steel reinforcement for structures of this type are selected.

$$L_d = \frac{V_h}{V_d} \times d_d \times f$$

Where:
L_d = settling length of the floating particle
V_h = horizontal speed at which the water should enter the settling basin
$0.2 < V_h < 0.4$ (m/s) is recommended
V_d = settling speed of the particle
d_d = settling depth (no more than 1.00 m is recommended)
f = safety factor for the settling length
$2 < f < 3$ is recommended

$$W = \frac{Q}{V_h \times d_d}$$

Where:
W = width of the settling basin (m)
V_h = horizontal speed at which the water should enter the settling basin (m/s)
d_d = settling depth (m)

In order to select the diameter of the sediment particles, take a water sample from the river near the intake, let it settle in a transparent container and observe or measure the size of the particles, from the smallest to the largest.

Comparing the coarse settling basin with the silt basin, the coarse settling basin sits deeper in the header tank as the gate remains open enough to expel thick material. On the other hand, V_h, is slightly more than 0.4 m/s because of the sedimentation of heavier particles larger than 1 mm in diameter and the V_d for those particles is more than 0.10 m/s. See Example 4 in Annexe I.

Table 4.1 Settling speed of the particles, depending on the diameters

Diameter of particle (mm)	Settling speed V_d (m/s)
0.1	0.02
0.3	0.03
0.5	0.1
1.0	0.4

COARSE SETTLING BASIN

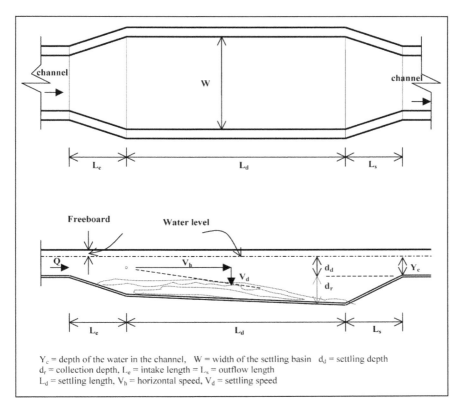

Y_c = depth of the water in the channel, W = width of the settling basin d_d = settling depth
d_r = collection depth, L_e = intake length = L_s = outflow length
L_d = settling length, V_h = horizontal speed, V_d = settling speed

Figure 4.2 Diagram of a settling basin

Construction procedure

The construction begins at the end of the headrace channel. The steps to be followed are as follows.

Construction of the platform

- A similar process to the construction of the headrace channel explained in Chapter 3.

Construction of the settling basin

- Draw and stake out the centre line of the settling basin on the platform, extending the centre line of the channel.
- Draw the borders of the settling basin. In this case they are the same borders as the channel, bearing in mind the total width and the thickness of the slopes = b + 2e.

- Dig the trench for the settling basin, bearing in mind the different depths d_1 and d_2 (in this case the same as the channel), including the breadth of the floor and the walls.
- As the excavation progresses, put stakes in place at the level required by d_1 and d_2 and the thickness of the floor.
- Shape the excavation and remove large stones from the dug out ditch.
- Verify the bearing capacity of the soil on the building site.

Waterproofing the settling basin

- Place 5 cm thick paving on the bottom of the excavated trench, bearing in mind the direction of the paving in accordance with depths d_1 and d_2.
- Draw the centre line of the settling basin on the paving as a guide for the rest of the layout.
- On the paving, complete the layout of the settling basin, determining the width and thickness of the walls in order to build the steel reinforcement structure and form the walls.
- Build and put the steel reinforcement in place, distributing it in accordance with the plan or the details of the structure on the floor and the walls (Figure 4.3).
- Put concrete blocks (paving separators) in place to give the steel the coating indicated in the plan.
- Put guiding points on the floor for the cement fill, at the indicated height, discounting the final coat.
- Prepare the concrete in accordance with the specified quality (f'_c) and pour it on the floor of the settling basin.
- Draw the central line of the settling basin again and the width and thickness of the walls as a guide for forming the walls.
- Plot the location of the side gate.
- Form the side walls, put separators between the vertical steel rods for the coating and fit the side gate in the correct position.
- Prepare the concrete and cast the walls to the indicated level, applying the casting technique to prevent holes, sweating, segregation, etc.
- Strip the form no earlier than 48 hours and cure them immediately.
- Plaster the floor and the walls, giving them a polished finish.
- Cure (submerged) for at least eight days.
- Test the gate to make sure it opens and closes properly.

Technical specifications

These refer to the quality of the concrete, the quality of the foundation soil, the quality of the steel reinforcement and the characteristics of the gate.
- $f'_c = 175$ kg/cm^2
- Dosage volume: 1:2:3 (cement:sand:stones)
- $f_y = 4,200$ kg/cm^2

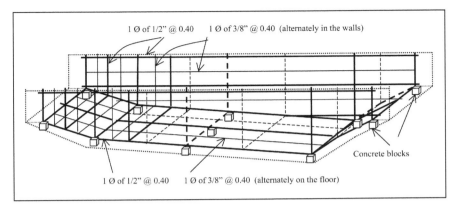

Figure 4.3 Assembly of prepared steel structure for the walls and floor of the settling basin

- Coating = minimum 3 cm
- Steel joints for 3/8" (1.0 cm) = minimum 0.20 m, for ½" (1.3 cm) = minimum 0.30 m
- μ_T = 1.5 kg/cm² (minimum)
- Submerged curing: At least eight days
- Inner plaster: 3:1 sand-cement, polished finish
- Wheel sluice gate, 0.25 wide by 0.20 high.

CHAPTER 5
Conveyance channel

The conveyance channel can be a natural or artificial channel in which the water flow is driven by gravity. It comprises the channel itself and complementary works.

In most cases, the water surface is exposed to the atmosphere. In some cases, the channel is said to be closed when piping is used, although not under pressure but partly filled.

The types of channel most frequently used in micro hydroelectric power plants, depending on the section, are shown in Figures 5.1 and 5.2.

The channel elements are:
- Geometric elements in accordance with the cross-section (Figure 5.3)
- Hydraulic characteristics of the channel (Figure 5.4).

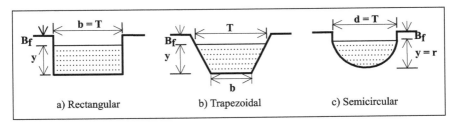

Figure 5.1 Open sections

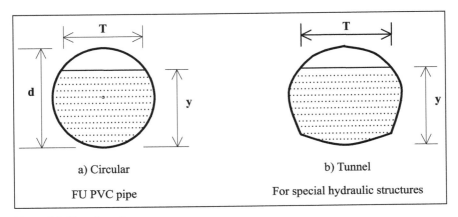

Figure 5.2 Closed sections

where:
b = Channel bed, T = Water surface, d = Diameter, y = Maximum water depth,
B_f = Freeboard

60 DESIGNING MINI AND MICRO HYDROPOWER SCHEMES

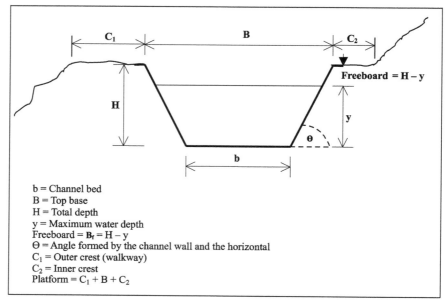

b = Channel bed
B = Top base
H = Total depth
y = Maximum water depth
Freeboard = B_f = H − y
Θ = Angle formed by the channel wall and the horizontal
C_1 = Outer crest (walkway)
C_2 = Inner crest
Platform = $C_1 + B + C_2$

Figure 5.3 Geometric elements of the channel section

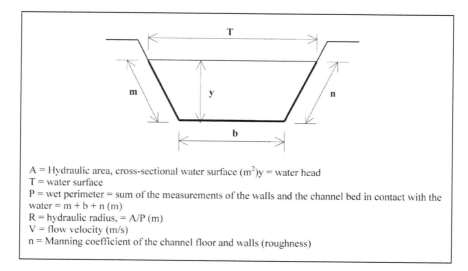

A = Hydraulic area, cross-sectional water surface (m^2) y = water head
T = water surface
P = wet perimeter = sum of the measurements of the walls and the channel bed in contact with the water = m + b + n (m)
R = hydraulic radius, = A/P (m)
V = flow velocity (m/s)
n = Manning coefficient of the channel floor and walls (roughness)

Figure 5.4 Hydraulic characteristics of the channel

The soil topographic conditions are good when:
- The gradient of the hillside forms a constant angle with the horizontal, no larger than 45°.
- The platform of the channel is stable throughout. In some cases, construction works such as aqueducts, bridge canals, etc. may be needed.
- There is a limited number of loose rocks on the hillside over the channel platform.

The stability of the soil is good when:
- There are no faults (settlements), landslides or subsidence, or if there are, they can be easily controlled with small complementary works.
- There is no trace of organic soil or soil of a plant origin, soil with a low bearing capacity, or soil that is swampy, sandy or of non-cohesive alluvial and slippery material.
- There are no stretches of expansive clay.
- The soil is uniform or has favourable small rocky segments that require no major investment costs involving explosives.
- There is no run-off during the rainy season, either towards the channel or that could be channelled through a storm aqueduct.
- There is no dampness or outpouring of water downstream or upstream from the channel.
- The majority is low permeability soil, so either no lining is needed or only some sections may require lining.
- No spillways are required to prevent unexpected subsidences, overflows or floods.

Among other factors, snow or frozen water during extreme weather conditions can affect the normal operation of the channel and the other civil works components.

Design of the channel (see Annexe Example 5)

The design of the channel consists of determining its dimensions for the safe passage of the designed water volume, with the minimum possible investment.

To this end, the following information is required:
- Formulas that help calculate the channel elements.
- Minimum and maximum speed tables, according to the type of material used, with or without lining.
- Determination of the slope (gradient), according to the type of soil or lining.
- Using maximum hydraulic efficiency channels is preferable.
- Determination of the section of the channel with a maximum hydraulic efficiency, based on the width of the platform.

Formulas used

1. Continuity formula

   $Q = A \times V$

   Where: Q = flow (m³/s)
   A = area of the cross-section of the channel (m²)
   V = speed of the water (m/s)

2. Manning formula:

   $V = \dfrac{1}{n} \times R^{2/3} \times S^{1/2}$

   Where: V = speed of the water (m/s)
   n = Manning coefficient of the lining or natural material
   R = hydraulic radius (m)
   S = channel slope (gradient)

3. Continuity and Manning formulas combined

   $Q = A \times \dfrac{1}{n} \times R^{2/3} \times S^{1/2}$

Table 5.1 Recommended wall slope for trapezoidal section channels

Material	Angle θ°
Sand	18.4
Sand and grit	26.6
Grit	34.0
Grit and clay	45.0
Clay	60.0
Concrete	60.0

Source: ITDG Peru, 1992

Table 5.2 Maximum flow velocity recommended

Material	Flow velocity (m/s) (depth less than 0.30 m)	Flow velocity (m/s) (depth less than 1.00 m)
Sand	0.3	0.5
Sandy grit	0.4	0.7
Grit	0.5	0.8
Clay grit	0.6	0.9
Clay	0.8	2.0
Masonry	1.5	2.0
Concrete	1.5	2.0

Source: ITDG Peru, 1992

Table 5.3 Manning coefficients

Type of conveyance surface	Manning coefficient: n
Earth channels:	
Clay	0.0130
Smooth solid material	0.0167
Sand with some clay or split rock	0.0200
Sand and gravel bottom and rocky sides	0.0213
Fine gravel 10 to 30 mm	0.0222
Regular gravel 20 to 60 mm	0.0250
Thick gravel 50 to 150 mm	0.0286
Grit in clumps	0.0333
Stone lining	0.0370
Sand, grit, gravel and grass	0.0455
Rock channels:	
Slightly irregular rock	0.0370
Irregular rock	0.0455
Very irregular rock with many outcrops	0.0588
Stone and cement masonry	0.0200
Masonry walls with sand and gravel base	0.0213
Concrete channels:	
Good cement finish (sealing coat)	0.0100
Gypsum or smooth concrete finish with a high content of cement	0.0118
Concrete with no sealing coat	0.0149
Concrete with a smooth surface	0.0161
Irregular concrete lining	0.0200
Irregular concrete surfaces	0.0200
Wooden channels:	
Planed and well-joined planks	0.0111
Rough planks	0.0125
Old wooden channels	0.0149
Pipes:	
Black commercial wrought iron pipes	0.012
Galvanized commercial wrought iron pipes	0.013
PVC pipes	0.010
Natural watercourses:	
Natural river bed with a solid bottom, no irregularities	0.0244
Natural river bed, with grass	0.0313
Natural river bed with stones and irregularities	0.0333
Torrent with irregular large stones, river bed with sediment	0.0385
Torrent with thick stones and plenty of sediment	0.0500

Source: ITDG, 1996

Open section channels with maximum hydraulic efficiency

These channels have a larger cross-section area and a smaller wet perimeter. They have a positive influence on excavation costs, given the lower volume of dug out material and the greater flow carried by this section. The most frequently used channels are shown in Figure 5.5.

Select one of these sections, depending on the platform and financial resources available for the channel.

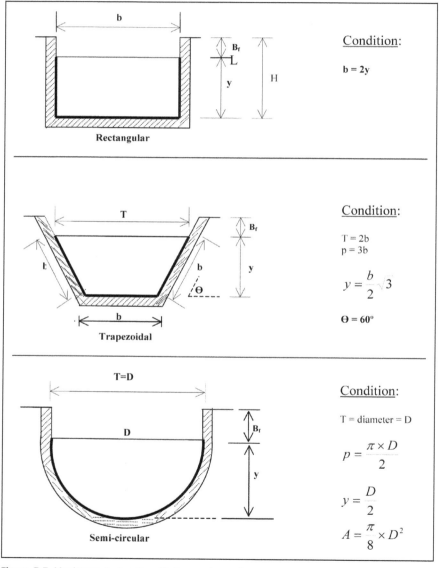

Figure 5.5 Maximum hydraulic efficiency channels

Closed circular section channel (see Annexe Example 6)

In some cases, when the flow is less than 0.100 m³/s and the gradients are less than 6.2 m/km, closed circular section channels made of flexible union (FU) PVC material have been conveniently used, as larger pipes that are non-standard and more costly are required for larger flows. Two cases are described here.

The first case consists of making the pipe work at a pressure greater than three-quarters of the inner diameter, but still operating part-full. For that condition, the formulas given in Figure 5.6 were generally applied, in which case the value of angle θ must be used correctly in radians and degrees.

The second consists of making the pipe work with a maximum depth value equivalent to three-quarters of the inner diameter (see Photograph 5.1). The formula shown in Figure 5.6 is used for a specific case, or following the manufacturer's recommendations, so that the PVC piping works as a channel rather than as a penstock. To obtain this formula, the Manning formula and the continuity law were applied and a value of $n = 0.01$ for the PVC case.

Advantages of this type of channel:
- anti-corrosive
- resistant to electrolytic corrosion
- lightweight
- economical
- easy to install and greater yield
- low maintenance and cost
- hermetic seal, guaranteed filtration-proof.

Photograph 5.1 Pipe to convey water where an open channel is impracticable

66 DESIGNING MINI AND MICRO HYDROPOWER SCHEMES

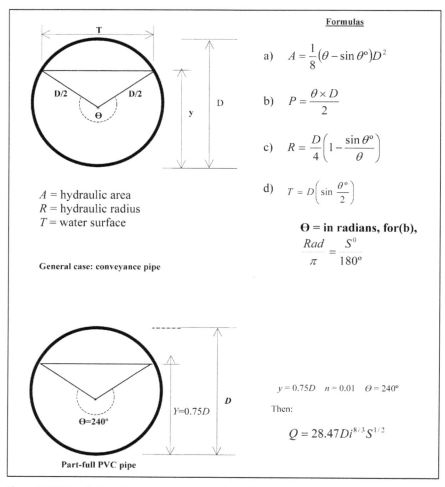

Figure 5.6 Manufacturer's recommendations for using piping as a channel

Disadvantages:
- Sometimes this type of channel may be not economical compared with concrete lining, especially when the required pipe size is not commercially available (in most cases this happens when the diameter exceeds 12" (30.5 cm). A economic comparison is recommended in order to ascertain the convenience of using this type of conveyance channel.
- They must be buried to protect them from the sun or extreme changes in temperature.
- When they change direction (bends) and for maintenance purposes, a curved channel with covers or concrete access points must be built, or curved PVC pipes must be manufactured in accordance with the radius of curvature.

Conveyance channel construction process

If the channel is built in earth or in rock cased with concrete, the steps to be followed are similar to those detailed for the construction of the headrace channel, with a few particular characteristics in each case. In general, these steps are as follows.

- Clear the land, lay out and stake out the platform for the channel and the gradient.
- Excavate the ditch.
- Apply a lean concrete bed and lining.
- Cure the concrete.
- Put the expansion joints in place.

In the case of PVC pipes, the excavation of the ditch should be supported by a layer of fine compacted material, making sure that the unions (socket joints) rest normally on the supporting base and are buried to protect them from the ultraviolet rays of the sun and impacts from stones that could cause a breakage.

Construction of a trapezoidal channel lining using the Practical Action wooden frame method referred to as 'cerchas'

This procedure has been tested in many channels for micro and mini hydroelectric plants in Peru and other countries, obtaining waterproof linings 2" to 3" (5.1–7.6 cm) thick. Instead of the conventional forming technology, wooden frames are used, which are referred to in Practical Action as 'cerchas'.

The results are economical and of good quality. Once the platform is built, the steps to be followed are outlined below.

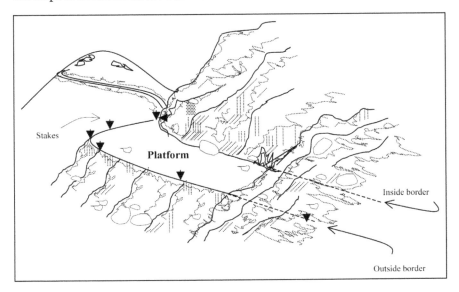

Figure 5.7 Open platform

Lay out and stake out the centre line of the channel (Figure 5.8)

- Draw a reference polygon formed by tangent lines at the intersection of the side slope and the platform.
- Take the measurements of interior crown C_2 and top base B from the corresponding plan.
- Parallel to the polygon drawn during the first step, put stakes in place at an equal distance to $C_2 + B/2$, every 10 m.
- Tie a rope around the tops of the stakes to determine the centre line and points of intersection, indicating a change of direction with lime or gypsum signs.
- Verify or stake out the centre line.

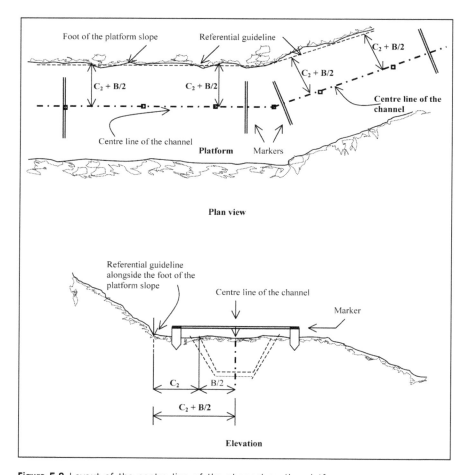

Figure 5.8 Layout of the centre line of the channel on the platform

Layout of the centre line in bends

This consists of obtaining a smooth and symmetrical curve, in accordance with the radius of curvature indicated in the plan.
- At each point of intersection (PI), take a distance equivalent to L, determined by points 'a' and 'b'.
- Draw line ab and measure distance d from the PI at the centre of line ab.
- Mark points 'c' and 'e', (half way between 'a' and PI and 'b' and PI respectively), then stake out the distances from these points to cord ab.

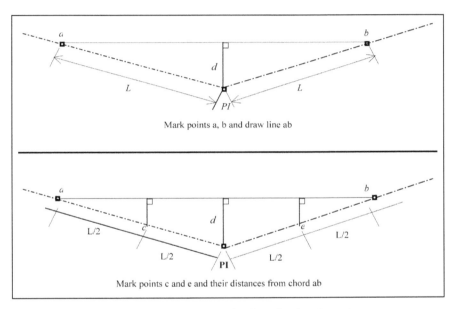

Figure 5.9 Marking points to draw curved border lines for the channel

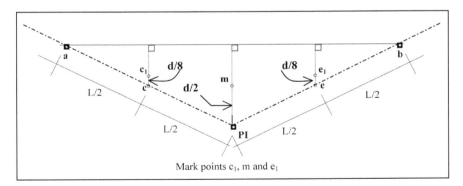

Figure 5.10 Marking points to draw curved border lines for the channel

70 DESIGNING MINI AND MICRO HYDROPOWER SCHEMES

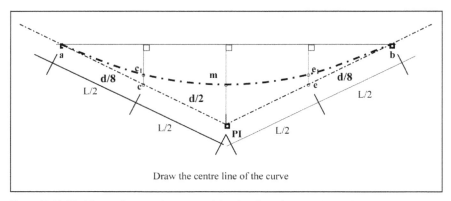

Figure 5.11 Marking points to draw curved border lines for the channel

- Mark points c_1, m, and e_1 in each of these distances, as follows:
  $c-c_1 = e-e_1 = d/8$; and $PI-m = d/2$
- With a rope, join points a, c_1, m, e_1 and b to obtain the centre line of the curve.

Layout of the ditch for the excavation work.

This consists of drawing the borders of the ditch on the platform, including the breadth of the covering.
- Taking the centre line drawn in both straight and curved sections as a main reference and using marking tools and material, draw the edge of the channel bed on both sides and at right angles with the centre line, measuring a distance = $b/2 + 0.58e$.
- Likewise, draw the borders at the top of the excavation, measuring $1 = B/2 + 1.16e$ on both sides.
- e = thickness of the walls and the floor, when $\theta = 60°$ (see Figure 5.12).

Excavation of the ditch (Figure 5.13)

This consists of the necessary excavation work to obtain the required trapezoidal section.
- Begin in the middle, as though it was a rectangular ditch limited by the edge determined by points P and Q on the platform and the channel bed equivalent to width: $b + 1.16e$.
- Once the total excavation depth $H + e$ is reached, continue shaping the inclination of the slopes ($\theta = 60°$), beginning with the total width of the borders at the top of the excavation determined by points R and S, to reach the border of the channel bed determined by points P' and Q'.
- It is important to verify the slope at the bottom of the excavation with a topographic level or level hose, at least every 10 m in straight sections and every 5 m or less in curved sections.

CONVEYANCE CHANNEL 71

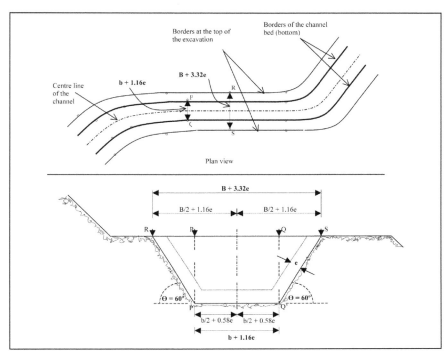

Figure 5.12 Layout of the borders of the channel for the excavation, on the platform

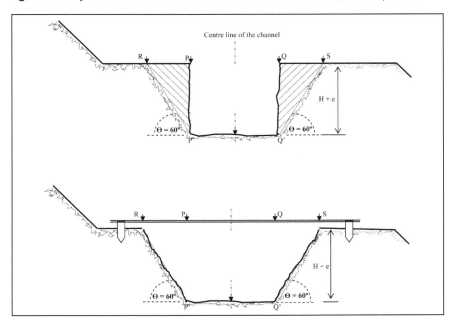

Figure 5.13 Excavation of the first and second parts of the trapezoidal ditch and moving points P, Q, R and S on the markers

72 DESIGNING MINI AND MICRO HYDROPOWER SCHEMES

Photograph 5.2 Excavating the ditch using a wooden frame.

Channel lining

Essential equipment and tools are:
- 3 m and 30 m measuring tapes;
- spirit level;
- cylindrical and conical plumb lines;
- framing square;
- rope;
- topographic level or level hose;
- stakes;
- 2″ (5 cm) nails;
- pick;
- spade and shovel;
- 3 to 4 pound (1–2 kg) muckle hammer (one end pointed);
- 16 ounce (450 g) hammer;
- plastering tool;
- stirring tool;
- mason's trowel;
- concrete tins;
- 1 wheelbarrow;
- wooden guide strip 1½″ (3.8 cm) × 3″ (7.6 cm) × 3.0 m;
- wooden frames the same as for the channel section with $\theta = 60°$ (Figure 5.14).

CONVEYANCE CHANNEL 73

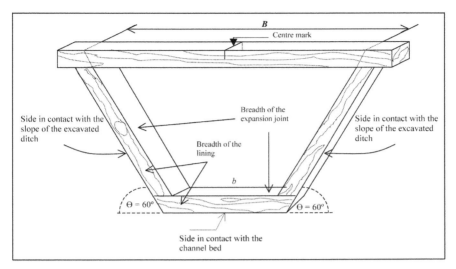

Figure 5.14 Details of the wooden frame

Lining consists of putting a layer of concrete of a specified resilience and uniform thickness on the ground and on the channel walls or slopes, giving it a polished finish with cement and fine sand and concluding with the expansion joints, following these steps:
Woodwork:
- Put stakes in place on the bottom of the channel bed, every 10 m on straight sections and every 5 m or less in curves, bearing in mind the design slope. Use the topographic level, tape measure, sight rods or level hose (Figure 5.15)

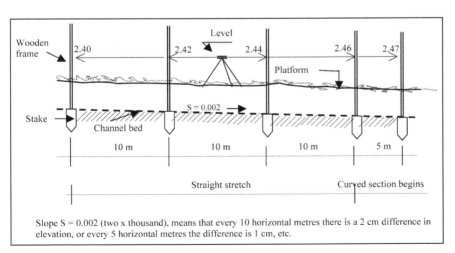

Slope S = 0.002 (two x thousand), means that every 10 horizontal metres there is a 2 cm difference in elevation, or every 5 horizontal metres the difference is 1 cm, etc.

Figure 5.15 Woodwork for preparing the channel bed

Installation of wooden frames:
- Place the master frames on the stakes.
- Align the master frames with the centre line of the channel to coincide with the taut rope between the markers and the marked centre of the frames.
- Plumb the master frames.
- Level the master frames with the spirit level.
- Fasten down the master frames by placing stakes on the slopes and tightening them with binding wire.
- Position intermediate stakes between the master frames, every 2.50 m.
- Put intermediate frames in place in the intermediate templates, making sure each of them is also aligned, plumbed and levelled.

Verify the slope, alignment, plumbing, levelling and clamping (Figure 5.16).

Lining and curing:
- Prepare the lining mix according to the specified dosage and materials. This is a semi-plastic mix (little water).
- First of all pour the mix on the slopes, smoothing, compacting and levelling with a wooden straight-edge, from one frame to the next. The thickness of the slope and the floor is controlled by the thickness of the frame.

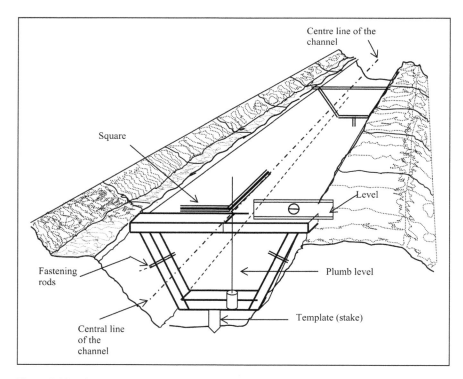

Figure 5.16 Alignment, levelling, plumbing and fastening of master frames

- Once the lining of the slopes is completed, remove the fastening rods from the frames and fill in with concrete.
- Pour the mix on the ground to form the channel bed, also compacting and levelling with a strickler.
- Whitewash the concrete surface of the slopes and the floor with a blend of cement and fine sand, in accordance with the technical specifications, using a plastering tool.
- Remove the frames after at least 24 hours, leaving the resulting spaces clean for the expansion joints.
- Cure the lining of the channel by completely covering it with water for at least 8 days. Curing guarantees the quality and durability of the lining. To fill with water, contain water in the lined channel by placing an earth dyke (small dam) at each end.

Filling in the expansion joints:
The expansion joints are determined by the grooves left by the frames when they are removed. These allow the concrete to expand or contract as a result of the temperature, preventing the concrete slabs from cracking and becoming permeable. Follow the steps below:
- Clear the grooves of any earth, gravel or other foreign matter with an adequate tool (pointing trowel), depending on the breadth and depth of the groove.
- Fill in the groove with the selected material (fine surfacing) and compact it to obtain a groove with a uniform width and depth.
- With a brush, prime the bottom surface of the groove with a tar and kerosene solution, as though it was paint.
- Fill in and compact the groove with a blend of hot tar and fine sand, trying not to spread it beyond the lined surface.

Note: This system has been used to line several channels carrying flows of 0.050 m^3/s to 1.5 m^3/s. The oldest one so far has been in service for 14 years.

As a result of the replication of this technology by trained people, normal yields have improved and the non-industrial manufacture of frames for flows greater than 0.500 m^3/s has been simplified.

Lining of a semicircular channel with the wooden frame method

The construction of semicircular channels with the conventional wooden forming method has been difficult, as it requires spending more on wood and manpower.

Designs or calculations of sections of maximum hydraulic efficiency channels show that semicircular sections are the cheapest. They require a narrower platform, less volume of concrete, less filling of expansion joints and less volume of excavation for the same flow and the same slope.

As an experiment, ITDG lined a 0.05 m thick circular section channel, using prefabricated frames with steel T profiles, in accordance with the project's geometric and hydraulic characteristics (see Figure 5.17).

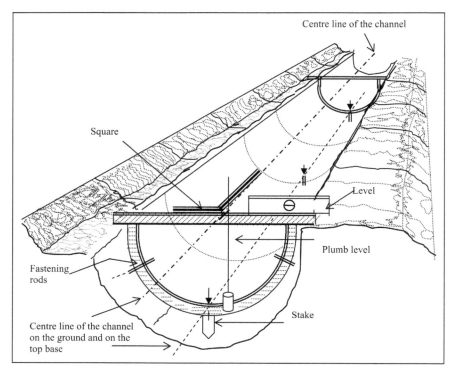

Figure 5.17 Alignment, levelling, plumbing and fastening of master frames, in a semicircular section

The steps to be followed are the same as for the trapezoidal section channel, adjusted to this type of section. The success lies in the layout and the excavation to depths no deeper than the sum of the radius and the free border. The woodwork, lining, curing and filling in the expansion joints are the same.

Technical specifications

The technical specifications refer to the geometric and hydraulic characteristics of the channel and the type of material, etc. When PVC pipes are used, the manufacturer's recommendations and specifications must be borne in mind. For example:
- design stress;
- resistance to movement;
- resistance to bending;
- resistance to compression;
- Young's modulus;
- friction or Manning coefficient, etc.

Works to complement the conveyance channel

When the topography of the land is difficult and problems occur during the construction of the conveyance channel, these should be dealt with in their entirety to guarantee the proper operation of the channel. Described below are some of the difficulties and the solutions proposed and executed, requiring complementary works.

Spillway

This subject was discussed in Chapter 3 as a side channel. In this chapter it is completed with the overflow channel that all spillways or side channels should have.

The location of the spillway was specified as follows:
- In a straight section of the headrace channel, for the purpose of evacuating the surplus flow during the high flow stage of the river.
- Alongside the conveyance channel where unexpected landslides could occur on the channel, to prevent the water carried by the channel from overflowing.
- In each settling basin and/or forebay tank, for maintenance purposes or to automatically evacuate the flow when the hydropower plant is stopped either by the operator for maintenance reasons or when it is stopped automatically because the control system activates.

The components of the spillway are the same as those of the conveyance channel, except that, due to the topography and soil stability and so as not to cause any erosion of the adjacent or surrounding soil, the components of one or the other may vary in the sense that they are not often all repeated; in other words, they are not fixed or standardized.

In many cases, the overflow channels have a regular or sharp slope and go through farmland, in which case the use of sewage pipes and flexible unions is appropriate because they are totally waterproof. They are buried in the soil at least 0.60 m deep and will not interfere with farming activities. In other cases, it is customary to use stone masonry with cement mortar and sand.

The design of the spillway or side channel was explained in Chapter 3. The design of the overflow channel consists of applying the formulae explained in the section 'Design of the channel' earlier in this chapter, to which end it is necessary to find out the slope, maximum speed limits, roughness values, etc. and determine the desired section.

It is worth mentioning that rocky soil with natural conditions for conveying water safely is very favourable for building the overflow channel. The designer must use his or her criteria to take advantage of the existing topographic resources in the area, without causing any harm to the environment and obtaining the consent of the rural population who live there.

The technical specifications of the spillway and the overflow channel will depend on the materials to be used (concrete, PVC pipes, high density polyethylene, masonry, etc.). Their purpose is to guarantee the quality, durability and construction method.

The construction process for the overflow channel is determined by the type of material selected. Here the designer should employ criteria based on the topography of the land and the project's economic situation. However, the ideal in all cases is that the spillway serves its purpose, preventing erosion, flooding or any other disasters on adjacent lands, through the overflow channel.

Aqueduct (Figure 5.16)

This is a construction that gives the channel continuity when its anticipated course is interrupted by a watercourse conveying water either permanently or during the rainy season.

Design characteristics and criteria

- Aqueducts are usually made of reinforced concrete. Sometimes the clear span is considerable, making intermediate girders necessary. Therefore, additional costs may occur; however it is expected that the structure should not make the project expensive.
- Either PVC pipes with wooden girders, or iron pipes can be used as aqueducts. In all cases, there should be no deflection and they should be rigid enough to withstand the force of the wind and other stresses that could jeopardize stability.
- They are straight when the platform span runs across the channel and curved when the channel changes direction in that span.
- It is advisable for the aqueduct to be a section of the conveyance channel, to avoid the additional construction of transition sections at the inlet and outlet points of the aqueduct.
- Sometimes these transition sections are unavoidable, when the conveyance channel has a trapezoidal section and the aqueduct needs to be rectangular because it is easier to build.
- In other cases, when the clear span is considerable and reinforced concrete is too expensive, then other design criteria are required, using materials that cost less, weigh less, but are safe. For example, in the El Tingo MHS, the direction of the channel had to be changed, with a 24 m clear span running from one side of the river to the other to carry 0.60 m^3/s. Three girders were designed, one at each end in natural soil and an intermediate one at 16 m from one of the ends. It was determined that the intermediate support should be a 2.50 m tall reinforced concrete column and the girders at the ends would be simple support beams. It was also established that the support beam in the 16 m section would be a construction steel lattice beam and in the remaining 8 m, a wooden beam. A semicircular section channel was built over these beams, with galvanized zinc sheeting to make it lighter, taking advantage of the steep slope formed by the uneven ground between one side of the river and the other.

CONVEYANCE CHANNEL 79

The steep gradient produced a high speed, although permissible for the zinc sheeting. To slow down the force of the water at the lowest end, a dissipation pond was built so that the water could then go through the conveyance channel under normal conditions (see Photograph 5.3). In another section, a type of hanging aqueduct suspended by a cable had to be built (Figure 5.18).

Photograph 5.3 Aqueduct supported by a steel lattice beam

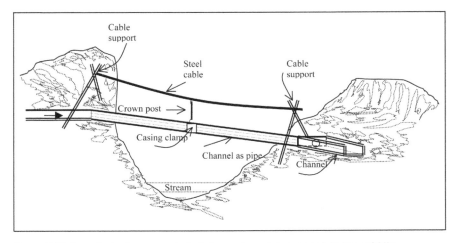

Figure 5.18 Aqueduct pipe supported at the ends and suspended in the middle

80 DESIGNING MINI AND MICRO HYDROPOWER SCHEMES

- An aqueduct in a curve with a trapezoidal section has its own particular characteristics with respect to the formwork of the channel bed and the sides. On the sides, the formwork is not done on a flat surface, but on a revolution surface, where the elements of a revolution cone must be determined (apex, element, etc. with the radius of curvature of the aqueduct). In this case, the formwork of the sides involves forming a strip on the side surface of the cone.
- An overhanging or balcony type channel also had to be built on a rocky slope with angles ranging from 45° to 65°, due to the topography of the land. Advantage was taken of the natural gradient for the sloping channel side and the channel bed and the other (vertical) side wall were completed with reinforced concrete (see Figure 5.19).

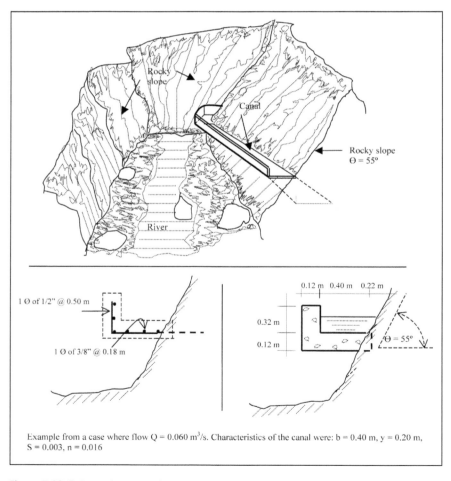

Figure 5.19 Balcony type canal

Storm aqueducts (Figure 5.20)

These are structures designed to direct run-off water over the channel to prevent further flow and debris from entering the channel. In some cases they also serve as passages over the channel from the bridle path, like bridges for pedestrians and working animals.

They can be made of simple concrete when the clear span on the top base of the channel does not exceed 0.60 m and should have a minimum thickness of 0.12 m. If the clear span is more than 0.60 m, then they should be made of reinforced concrete.

Lateral pressure-breaking drains (Figure 5.21)

In one section of the channel put into operation in the Huacataz MHS in Cajamarca, Peru (1991), two of the wall sections (sides) of the channel cracked as a result of the lateral pressure of the soil saturated with water. The water springs out during the rainy season when the water table rises.

For the repair work, it was decided to eliminate the water pressure with a drain, digging out and removing the sandy-clay soil in contact with the channel wall and replacing it with granular material (gravel). In addition, 2" (5.1 cm) diameter PVC pipes were embedded in the slope, so that the water would drain towards the channel. It was determined that the increased flow in the channel was manageable, as the drained water reached 1.5 litres/s only during the rainy season.

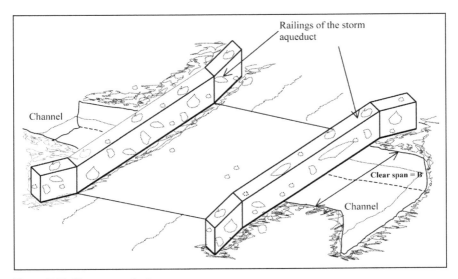

Figure 5.20 Storm aqueduct

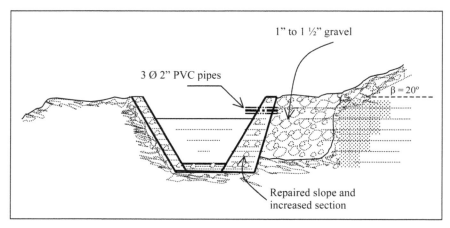

Figure 5.21 Pressure-breaking drain

Platform wedging (Figure 5.22)

These are artificially built structures to piece together or complete the width of the channel platform.

Stones and clay soil are used for the construction. The stones should preferably be angular with flat sides. Sand or clay is usually found in highland areas. In jungle fringe areas, on the other hand, cobbles or boulders found in the river can also be used for wedging.

In some channel sections, these structures can be as tall as 0.80 m. Local farmers are familiar with the technology because they use it to build fences around their properties.

Diversion channels or gutters (Figure 5.23)

- These are located on the slope upstream and nearly parallel to the conveyance channel.
- They resemble a collecting channel that receives run-off from the slope (during the rainy season) to carry it adequately towards the natural drains or watercourses, without causing any harm to the surrounding land.

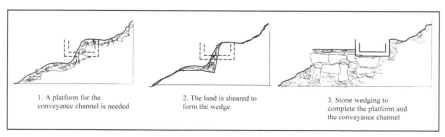

Figure 5.22 Stone wedging for the platform of the conveyance channel

- They prevent the run-off and debris from entering the conveyance channel.
- They are also used to protect road drains and road platforms, or any other lower level installations.

For the design of diversion channels, it is advisable to obtain the advice of a soil conservation expert, because it is not a matter of protecting only the conveyance channel, but all the land on the slope and in the mini or micro watershed, which usually includes farmland, natural pastures for grazing cattle, fallow land in continuous erosion, etc.

The expert should provide essential elements for calculating these channels. For example, the maximum design run-off for a return period of at least 10 years, is calculated using the following formula:

$$Q = \frac{C \times I \times A}{360}$$

where:
Q = discharge (m³/s)
C = run-off coefficient, depending on the relief, texture, plant cover conditions, etc.
I = maximum rainfall intensity during a specific time, equivalent to the concentration time (T_c) of the catchment, expressed in mm/h
A = catchment area, in hectares.

This formula is used to calculate the maximum peak discharge for a specific frequency period, which is very important for designing the elements for diverting surface water, such as diversion channels or gutters, sewers, rapids, falls, drains, etc.

There are specific methods for calculating each of the parameters C, I and A, each depending on other parameters.

For example, concentration time T_c is the time it takes the water to travel from the most remote point in the catchment area to the discharge or control point, assuming that the distance travelled by the water is equivalent to the length of the drainage course or river, which can be estimated based on the gradient and length of the course of the river, the average slope of the land and its plant cover, or based on nomographs,.

Furthermore, rain precipitation records need to be obtained from strategically located rain gauging stations.

In the case of micro hydropower schemes to be established in isolated areas where there are no useful records and/or studies for complementary works of this type to protect the conveyance channel, engineers should resort to the following options.

- Reach an agreement with the owners of the land adjacent to the conveyance channel and give them the technical guidelines and facilities for them to build the diversion channel in the required areas.
- In other cases it was observed that, in some sections, the soil on the slope over the channel could slide, not because of the run-off during the rainy

season, but because of uncontrolled practice of gravity-fed irrigation. In this case, arrangements were made with other development units in the rural area to introduce technical irrigation practices (sprinkler irrigation) or other appropriate technologies (reforestation).

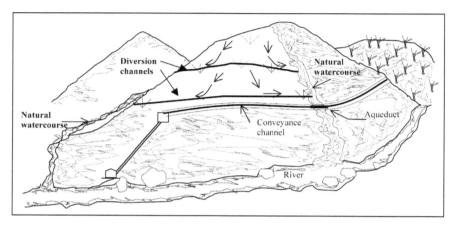

Figure 5.23 Diversion channels

CHAPTER 6
Silt basin

The silt basin is a structure in which the solid particles floating in the water leading to the conveyance channel can settle and then be evacuated to an appropriate place. It is located:
- At the end of the headrace channel (when there is no settling basin).
- At the end of the conveyance channel (next to the forebay tank).
- In some cases halfway, or in a strategic intermediate point in the conveyance channel when the channel is fairly long, e.g. more than 400 m, or when run-off enters the channel.
- On stable soil platforms, where the admissible bearing capacity of the foundation soil is greater than the stress transmitted by the weight of the entire structure.
- In a wide enough area not prone to subsidence or landslides in adjoining areas.
- The bottom level must always be higher than the level of the overflow channel and the level of the latter must always be higher than the high flow level of the river or than the level of the land or area in which the sediment material will be evacuated.
- Preferably on a straight stretch, with the centre line in line with the headrace channel or the conveyance channel.

The geometric shape, elements and complementary works are shown in Figures 6.1 and 6.2.

Description and characteristics of the silt basin elements

L_e = intake length

- Similar to a transition area with a larger cross-section than the conveyance channel.
- Variable widths (B) and depths (y)
  B, increases in value, to W.
  y, increases in value, to d_1.
- In this stretch, the dimensions of the cross-section area vary so that the speed (m/s) of the water entering the channel decreases in the silt basin to a V_h value.

86 DESIGNING MINI AND MICRO HYDROPOWER SCHEMES

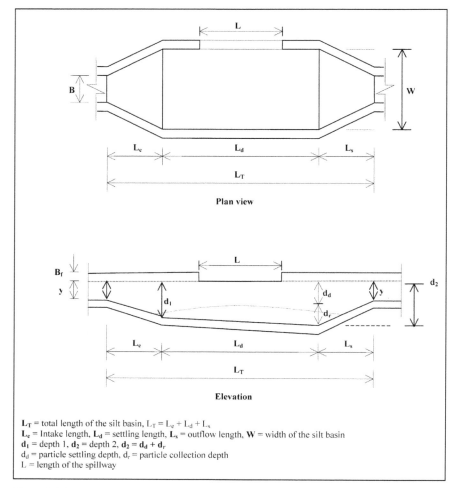

Figure 6.1 Silt basin elements
Note: Elements of the conveyance channel: B = top base, y = water head, B_f = freeboard. In Figure 6.1, overflow length L_T was added.

L_d = settling length

- The area where the particles are decanted, with horizontal speed V_h descending to vertical distance d_d with settling speed V_d
- This section has a constant width W and its depth varies from d_1 to d_2.

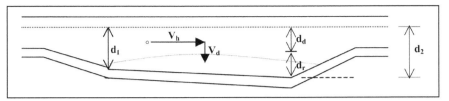

Figure 6.2 Vertical section view of the silting tank

L_s = outflow length

- A transition area similar to L_e, joining the silt basin to the channel.
- In this stretch, the cross-section of the silt basin varies from a maximum to a minimum, so that the water entering the channel recovers its speed and continues carrying flow Q without any suspended solid particles. W is reduced to the value of B

  d_2 is reduced to the value of y

  d_1 = depth 1
- Greater than the head, gradually increasing the cross-section area of the silt basin.

  d_2 = depth 2
- The deepest part of the silt basin which bears a relation to L_d, to ease the evacuation of the accumulated sediment during cleaning.

  W = width of the silt basin
- Corresponds to the L_d area and is greater than B of the channel, to create a larger area with d_d and obtain a V_h equivalent to the selected value.

  d_d = particle settling depth
- It is recommended that this depth is either the same as or slightly deeper than the head.

  d_r = depth of the collection tank or sediment material

  V_h = horizontal speed of the water in the settling area
- Recommended value: $0.2 < V_h < 0.4$ m/s

  V_d = particle settling speed
- The value is selected in accordance with the diameter of the smallest particle to be sedimented in the silt basin, in accordance with Table 6.1.

Table 6.1 Recommended speeds to settle different sizes of particles

Diameter of particle (mm)	Settling speed V_d (m/s)
0.1	0.02
0.3	0.03
0.5	0.1
1.0	0.4

Source: Harvey (1993)

Works to complement the silt basin include:
- Spillway or side channel to evacuate the excess flow or the total flow. This is located alongside the conveyance channel and the silt basin.
- Overflow channel: this receives the water from the spillway and conveys it to an adequate area without causing any harm. It may be made of concrete, stone masonry or a mixture of these materials, with PVC pipes.
- Cleaning system: this evacuates the sediment through a bottom side gate leading to the overflow channel and/or silt basin. In other cases a drainage system with PVC pipes has been built underneath the bottom of the silt basin, connected to the overflow channel or silt basin with a piece of pipe as a stopper or sluice gate.

Design of the silt basin and complementary works (see Annexe Example 7)

1. Determination of W
   Begin with: $Q = A \times V_h$

   where:
   Q = flow (m^3/s)
   A = cross-sectional area of the silt basin (m^2)
   V_h = horizontal speed of the water (m/s).
   Previously calculate $A = W \times d_d$

   When replacing these values in the Q equation and clearing W, the equation obtained is:

   $$W = \frac{Q}{V_h \times d_d}$$

2. Calculation of L_d

   $$L_d = \frac{V_h}{V_d} \times D_d \times f$$

   where:
   V_h = horizontal speed of the water (more than 0.2 and less than 0.4 m/s)
   V_d = settling speed, depending on the diameter of the selected particle, in m/s (Table 6.1)
   d_d = sedimentation depth in m, assuming it is the same as the head of channel y, plus 0.05 m (approximately)
   f = safety factor ($2 < f < 3$)

3. Calculation of d_1
   Assuming that: $d_1 \geq 1.5y$

4. Calculation of d_2
   Assuming that: $d_2 = d_1 + $ (3% to 5%) L_d

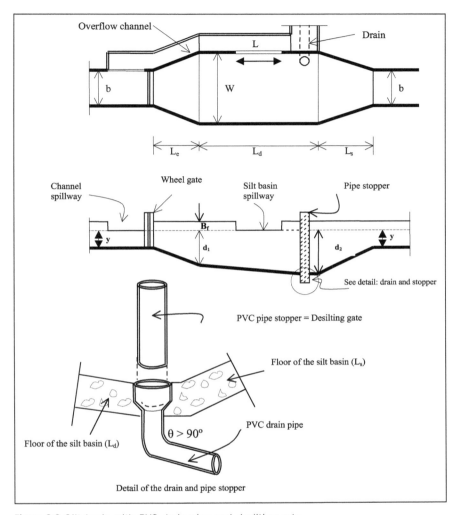

Figure 6.3 Silt basin with PVC drain pipe and desilting gate

For silt basins with flows less than or equivalent to 0.300 m³/s, PVC pipes were used to drain the sediment and the side wheel gate was replaced by a pipe stopper. Figure 6.3 illustrates this method. PVC is a low cost alternative and has been used extensively by the authors with excellent results.

Silt basin construction process

The following are the typical construction steps for areas where the soil is not difficult to excavate, the use of explosives is not necessary and the bearing capacity is adequate.

90　DESIGNING MINI AND MICRO HYDROPOWER SCHEMES

- Clear away material from the area to be used:
  the section of the conveyance channel with a spillway and head gate;
  the silt basin;
  the overflow channel;
  a continuous section of the conveyance channel.

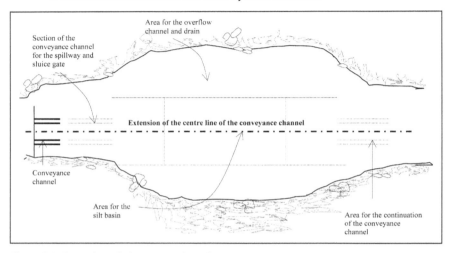

Figure 6.4 Extension of the centre line of the channel on the prepared platform

Photograph 6.1 Layout of the silt basin

- Shear and/or level the land for the platform on which the silt basin and its complementary works will be built.
- Draw and extend the centre line of the silt basin, using the centre line of the conveyance channel as a reference (Figure 6.4 and photograph 6.1).
- Plot the measurements of the silt basin with the excavation measurements of W, including the thickness of the walls, and mark L_e, L_d, L_s, the transition area and the width of the channel, including its thickness (Figure 6.5).
- Put level reference points in place to indicate the depth of the excavation, placing stakeout rods on either side of the central line, in pairs, at right angles with the centre line (Figure 6.6).
- Dig the first layer of the silt basin tank until depth is:
  $(B_f + y + e)$ initially, and
  $(B_f + y + e + S \times L_T)$ at the end (S = channel slope) (see Figure 6.7).
- Continue digging to the foundation levels, forming the edges or ramps for the floor and shaping the slope. The foundation levels in natural ground are: (see Figure 6.8):

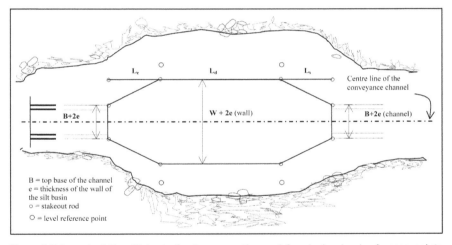

Figure 6.5 Layout of the silt basin for the excavation and for placing level reference points

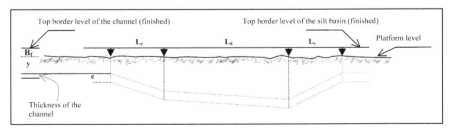

Figure 6.6 Level reference points for the excavation of the silt basin tank

92 DESIGNING MINI AND MICRO HYDROPOWER SCHEMES

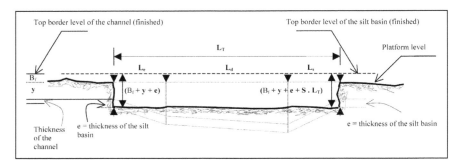

Figure 6.7 First part of the excavation of the silt basin tank, from the top border of the silt basin until depth: $(B_f + y + e)$ and $(B_f + y + e + S.L_T)$

$(B_f + y + e)$, when connecting with the channel or at the beginning of L_e
$(B_f + d_1 + e)$ at the end of L_e, and beginning of L_d
$(B_f + d_2 + e)$ at the end of L_d, and beginning of L_s
$(B_f + y + e) + S \times L_d$ at the end of L_s, connecting with the continuous channel.
- Complete the excavation of the silt basin tank (see Figure 6.9).

Note: Dimension *e* refers to the thickness of the floor; approximately 0.075 m to 0.10 m can be added for a layer of paving to obtain a uniform and accurate layout on the floor slab. If the soil is not suitable for a foundation, the best alternative must be determined on site to improve its bearing capacity.
- Stake out and put level points in place on the floor.
- Prepare the right proportion of materials and pour on the layer of paving up to the indicated levels, smoothing it and forming the ramps on the floor of L_e, L_d and L_s.
- Draw the centre line of the silt basin on the floor slab, using a plumb, nails, markers, rope, etc.

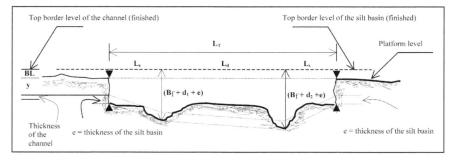

Figure 6.8 Location of the foundation levels, where the ramps on the floor of the silt basin change direction (the excavation for floor slabs is optional, depending on the quality of the soil)

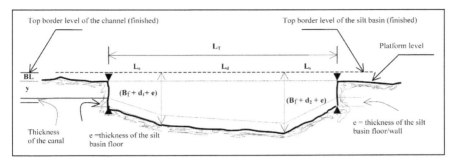

Figure 6.9 Total excavation of the silt basin tank

- On the floor slab, draw the measurements of the silt basin and leave guidelines for the distribution of reinforcement steel that will be laid lengthwise and crosswise on the floor.
- Prepare the steel reinforcements and distribute them in accordance with the technical specifications for distance, fold, diameter, etc. indicated in the structural plan (Figure 6.10 and Photograph 6.2).
- The crosswise steel bars that also form part of the walls must remain in a vertical position, braced with lengthwise steel bars and clamped down with steel remnants nailed to the natural slope of the silt basin tank.
- Prepare the concrete to be poured on the floor as near as possible to the silt basin, in the specified proportions.
- Place the casting levels on the ground, bearing in mind the thickness of the final coating.
- Lay the floor up to the indicated levels, stirring and smoothing the concrete with concrete tins, preventing it from segregating or sweating and compacting it appropriately; it should neither be porous nor holey.
- Clean the concrete residues that may be adhered to the crosswise steels.
- Draw the centre line of the silt basin on the concrete floor (after at least 24 hours) in order to draw the forming guidelines, discounting the coating (Figure 6.11).

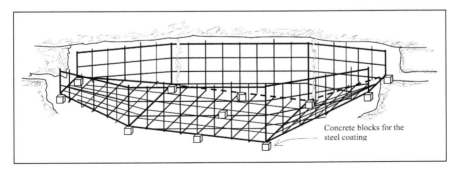

Figure 6.10 Assembly of the steel reinforcement resting on concrete blocks in the silt basin

94 DESIGNING MINI AND MICRO HYDROPOWER SCHEMES

Photograph 6.2 Placing steel reinforcements in the silt basin

- Finish putting the crosswise steel bars in place in the walls, in accordance with the specified spacing.
- Draw the central line and the level of the wheel gate and fit it before the general forming work.
- Form the outlet window and the remaining components of the wheel gate.

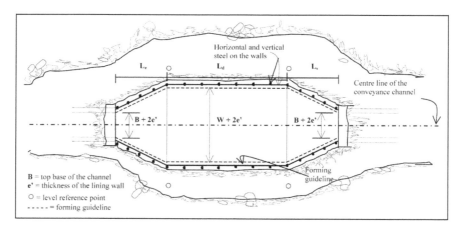

Figure 6.11 Layout of the centre line of the silt basin on the floor and forming guidelines for the walls

- Form the walls of the silt basin and the spillway, applying the forming technology (separators, bracing struts, etc.).
- Put height levels in place for lining the walls and the side spillway channel.
- Dampen the insides of the formwork and clean the base of the perimeter walls with water.
- Pour the concrete in accordance with the technical specifications, using standard practices (smoothing, compacting, etc.).
- Strip the forms after at least 24 hours in climates with a minimum daytime temperature of 12°C.
- Check the operation of the wheel gate.
- Plaster the walls and floor with the specified sand cement mortar.
- Cure the silt basin, making sure it is submerged (underwater).

These steps can also be applied for the construction of the channel spillway, the sluice gate and the overflow channel, bearing in mind their dimensions, slope, unevenness, the junction with the desilting channel, the expansion joint between the channel and the silt basin and their technical specifications, etc.

If PVC piping is used as a gate (stopper pipe) and for the drainage system, it is recommended that it should be no larger than 12" (30.5 cm), class 5, as this makes it easier to clean the silt basin. Larger diameters make maintenance operations more difficult and require other accessories.

The PVC drainage system requires a non-commercial elbow pipe larger than 90° to obtain the desired slope, so that the water carrying mud or sediment can be easily expelled. In practice, the elbow pipe was home-made to give it the angle required for the slope.

For example, if the drainpipe that will be joined to the homemade elbow pipe needs a 5 per cent slope, the elbow pipe must have a 95° angle (see Figure 6.12), as follows:

1. Place a bell-mouthed pipe horizontally on a flat surface.
2. Draw the 95° angle on the surface of the pipe, at a distance from the bell-mouth that includes the thickness of the floor of the silt basin as a minimum (built-in).
3. With due precision, shear the pipe with a saw.
4. Turn the piece of pipe without the bell-mouth and stick the sawn borders together to form the 95° bend.
5. On the 95° elbow pipe, mark points in pairs, one centimetre from the borders and every 4 cm, following the perimeter of the glued borders.
6. On the marks on both parts, drill the holes with a hand drill and a bit 1 mm larger than the diameter of the binding wire.
7. Join the two parts by stitching them together with binding wire.
8. The elbow pipe must be fitted into the floor of the silt basin at the same time as the steel, making sure that the bell-mouthed edge coincides with the level of the finished floor. It must also be levelled and plumbed.
9. The required length of the same pipe is used as a removable stopper.

96 DESIGNING MINI AND MICRO HYDROPOWER SCHEMES

Technical specifications

Silt basin:
- $f'_c = 175$ kg/cm² (cement: sand: pebbles = 1:2:3)
- $f_y = 4,200$ kg/cm²
- $r = 0.04$ m (floor and walls)
- $\sigma_T = 1$ kg/cm² (minimum)
- layer of paving, Cement: Aggregate, 1:12
- coating: cement sand, 1:3, polished finish

Spillway and overflow channel:
- rounded crest
- overflow channel : $f'_c = 140$ kg/cm², (cement: sand: pebbles = 1:2:4).

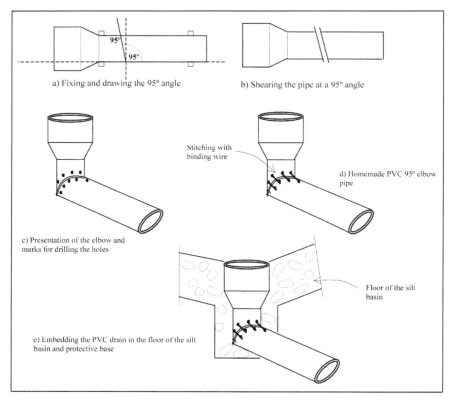

Figure 6.12 Homemade PVC elbow pipe for draining the silt basin using sections of pippes

CHAPTER 7
Forebay tank

The forebay tank is located at the end of the conveyance channel next to the penstock. Generally the silt basin is found immediately before the forebay tank. Its main characteristics are as follows.

- It allows the water to flow through to the penstock without any solid particles or other low density water components.
- It is strong enough to withstand the rush of kinetic energy of the water when there is a sudden change in the flow speed (excess water pressure in the penstock and the forebay, when the turbine comes to a sudden stop).
- It releases the excess flow or the total flow of the turbine through the spillway channel when it is necessary to switch off one or all of the injector valves in the powerhouse.
- It produces a constant head of water, from the level in the component considered, appropriate to the type of turbine selected for the project.
- It stops air from entering the pressure pipe.

The forebay is shaped like a tank (rectangular prism), the size of which is consistent with the technical recommendations. Its elements are as follows.

- Internal dimensions: length, width and depth (Figures 7.1 and 7.2). The length and width can be the same as the silt basin.
  Depth d_3 is measured from the water surface to the bottom:

  $d_3 = 4.5D + d$

  where:
  D = Diameter of the penstock
  d = Bevel (height from floor level to the lower edge of the penstock)

  This bevel provides space for the tapered inlet pipe entering the penstock and will prevent the operator from dropping a tool directly into the pipe during maintenance work.
  The total depth, d_T, of the forebay, from the roof to the floor, is

  $d_T : B_f + d_3$

- Tapered inlet pipe to the penstock: This is a reduction zone in the shape of a blunt cone. Its outside diameter (flush with the wall) is reduced to the diameter of the penstock. Its purpose is to ease the flow of water entering the pipe with a smooth contraction, thus preventing any great loss of head.
- Screen and roof: The screen is made of steel. It prevents suspended elements (pieces of wood, leaves, branches, plastic containers, etc.) from entering the pipe. It rests on and is anchored to the first wall that separates

the silt basin from the forebay tank and it is recommended that it forms a 60° angle with the horizontal to make it easier to clean the retained material.

It consists of a metal frame with bevelled section steel profiles; the distance separating them from each other must be smaller than the diameter of the injector for Pelton turbines with one injector, or smaller than the diameter of the smaller injector (when a multiple injector Pelton turbine is used); and no more than 2 cm in the case of other turbines such as the Michell Banki, Francis, Kaplan, etc. Otherwise, the manufacturer or supplier of the turbine should be consulted.

It is recommended that the profile of the lower part of the screen frame should have a circular section with two or three grooves for the clamps that will be fixed to the wall with anchored bolts that will act as hinges.

The top part of the frame leans on the edge of the roof. A padlock should be placed at this point to prevent it from being opened by irresponsible people who could cause serious damage to the turbine or the penstock.

The fixed roof may be made of concrete or strong timber to prevent debris from directly entering the forebay tank and/or penstock.

- Draining or cleaning system: Optional. In most cases cleaning is done manually, as the sediment could contain fine particles (lime, clay or fine sand). For easier cleaning purposes, a drainage system with PVC pipes connected to the spillway channel is advisable and a manhole with a screw top at floor level.
- Spillway and overflow channel: In most cases the forebay tank goes together with the last silt basin, in those cases one spillway is used for both (forebay tank and silt basin) otherwise the system should have its own spillway and overflow channel.

Design of the forebay tank (see Annexe Example 8)

This consists of establishing the dimensions and the quality of the concrete, reinforcement steel and complementary elements: fitting of the penstock, screen, cleaning system, etc.

The design is for a tank alongside the silt basin.

Determining the dimensions

Calculating the length: A value the same as L_s of the silt basin was assumed.
Calculating the width: A value similar to or less than width W of the silt basin was assumed.
Calculating the depth: It is very important to know diameter D of the penstock to be fitted into the forebay tank. Table 7.1 shows the referential inner diameter for PVC high pressure piping with flexible unions, bearing in mind moderate flow

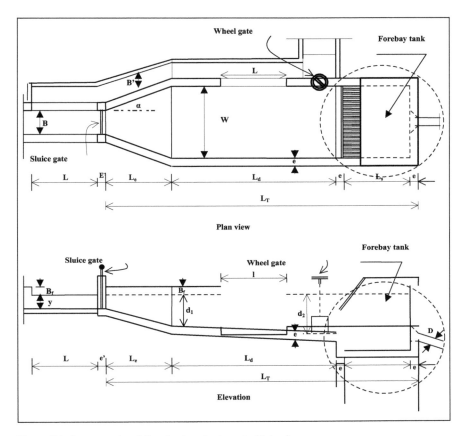

Figure 7.1 Components of the forebay tank and silt basin

speed hence moderate losses in the penstock. This table is useful to make preliminary estimations and for reference, since the actual penstock diameter must be determined using the equations and criteria in Chapter 8.

It is recommended that the head, from the surface of the water to the middle of the diameter of the penstock entering the forebay tank, should be at least 4D to ensure that there is a column of water above the mouth of the pipe inside the tank that will prevent air from entering the pipe. Also, any sudden change of pressure or any other interruption of the water flow inside the pipe can be dissipated or absorbed by that column of water and by the surge tank connected to the penstock before it is fitted into the forebay tank.

Based on this recommendation, it is determined that $d_3 = 4.5D + d$
The total inner depth is $d_T = d_3 + B_f$
d = bevel
The value of d is recommended to be at least 15 cm for small tanks where the design flow is below 0.050 m³/s.

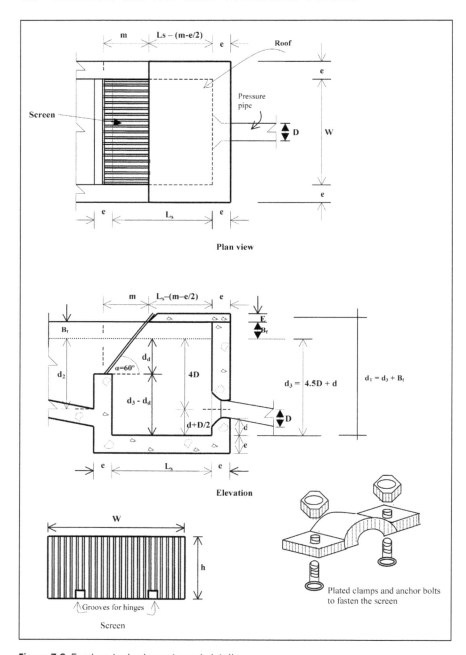

Figure 7.2 Forebay tank elements and details

Table 7.1 Range of flows for different internal diameters of penstock

Flow: Q (m³/s)	Inner diameter (referential): D_i (mm)
0.010–0.020	112.83
0.021–0.030	138.23
0.031–0.060	195.49
0.061–0.080	225.73
0.081–0.100	252.37
0.101–0.120	276.46
0.121–0.150	309.09
0.151–0.180	338.51
0.181–0.200	356.82
0.201–0.250	398.94
0.251–0.300	437.01
0.301–0.400	504.62
0.401–0.500	564.19
0.501–0.600	618.19

Source: Author's experience

For higher flows, the value of *d* is based on the diameter of the penstock and its tapered inlet pipe, between 15 and 50 cm.

The height of the common wall between the forebay tank and the silt basin is determined by values d_3 and d_d.

Calculating the roof slab: The inner width is the clear span provided by W and the outer width = W + 2e

The other length of the roof is determined by $L_{int} = L_s - (m - e/2)$ and the outer length is $L_{ext} = L_s + 1.5e - m$ (see Figure 7.2).

Reinforcement steel in the silt basin and forebay tank

This depends on the quality of the concrete, the bearing capacity of the soil and the pressures involved, such as the actual weight, the weight of the water, thrust of the soil, dynamic loading, etc.

Tables 7.3 and 7.4 help to determine the steel reinforcement required for both the silt basin and the forebay tank. These tables are for typical cases, as the design parameters used were f'_c = 175 kg/cm² of concrete, soil bearing capacity of no less than 1.5 kg/cm², different design flows and calculated dimensions.

For the roof of a forebay tank measuring approximately 0.60 m × 0.42 m including the breadth of the supporting wall, the concrete cover must be removable for maintenance operations. The roof slab should be at least 0.08 m thick, reinforced with ¼" (0.64 cm) steel both ways, separated by 0.15 m both ways. Otherwise, it could be replaced by a wooden roof.

Table 7.4 was prepared for larger than the above-mentioned dimensions, considering the measurements in Tables 7.2 and 7.3.

Table 7.2 Calculations for the components of the channel, silt basin and forebay tank

	Components of the rectangular conveyance channel													Components of the silt basin and forebay, together									
y	b	A	p	R	R*2/3	n	1/n	S	S*1/2	V	Q	V_h	d_d	W= $L_e=L_s$	60%Y	d_1	V_d	L_d	$0.04L_d$	d_2	D_i	d	d_3
m	m	m²	m	m						m/s	m³/s	m/s	m	m	M	m	m/s	m	m	m	m	m	m
0.10	0.20	0.02	0.40	0.05	0.14	0.014	71.43	0.002	0.045	0.44	0.009	0.3	0.15	0.20	0.06	0.16	0.03	3.00	0.12	0.28	0.075	0.15	0.49
0.12	0.24	0.03	0.48	0.06	0.16	0.014	71.43	0.002	0.045	0.50	0.014	0.3	0.17	0.28	0.07	0.19	0.03	3.40	0.14	0.33	0.096	0.15	0.58
0.14	0.28	0.04	0.56	0.07	0.17	0.014	71.43	0.002	0.045	0.55	0.022	0.3	0.19	0.38	0.08	0.22	0.03	3.80	0.15	0.38	0.117	0.15	0.68
0.15	0.30	0.05	0.60	0.08	0.18	0.014	71.43	0.002	0.045	0.58	0.026	0.3	0.20	0.43	0.09	0.24	0.03	4.00	0.16	0.40	0.129	0.15	0.73
0.16	0.32	0.05	0.64	0.08	0.19	0.014	71.43	0.002	0.045	0.60	0.031	0.3	0.21	0.49	0.10	0.26	0.03	4.20	0.17	0.42	0.140	0.15	0.78
0.18	0.36	0.06	0.72	0.09	0.20	0.014	71.43	0.002	0.045	0.65	0.042	0.3	0.23	0.61	0.11	0.29	0.03	4.60	0.18	0.47	0.164	0.20	0.94
0.20	0.40	0.08	0.80	0.10	0.22	0.014	71.43	0.002	0.045	0.70	0.056	0.3	0.25	0.75	0.12	0.32	0.03	5.00	0.20	0.52	0.189	0.20	1.05
0.25	0.50	0.13	1.00	0.13	0.25	0.014	71.43	0.002	0.045	0.81	0.101	0.3	0.30	1.12	0.15	0.40	0.03	6.00	0.24	0.64	0.254	0.20	1.34
0.30	0.60	0.18	1.20	0.15	0.29	0.014	71.43	0.002	0.045	0.91	0.164	0.3	0.35	1.57	0.18	0.48	0.03	7.00	0.28	0.76	0.323	0.25	1.71
0.35	0.70	0.25	1.40	0.18	0.32	0.014	71.43	0.002	0.045	1.01	0.248	0.3	0.40	2.06	0.21	0.56	0.03	8.00	0.32	0.88	0.397	0.30	2.09
0.40	0.80	0.32	1.60	0.20	0.35	0.014	71.43	0.002	0.045	1.10	0.353	0.3	0.45	2.62	0.24	0.64	0.03	9.00	0.36	1.00	0.474	0.35	2.48
0.45	0.90	0.41	1.80	0.23	0.37	0.014	71.43	0.002	0.045	1.19	0.483	0.3	0.50	3.22	0.27	0.72	0.03	10.00	0.40	1.12	0.555	0.40	2.90
0.48	0.96	0.46	1.92	0.24	0.39	0.016	62.50	0.002	0.045	1.09	0.502	0.3	0.53	3.16	0.29	0.77	0.03	10.60	0.42	1.19	0.565	0.50	3.04
0.50	1.00	0.50	2.00	0.25	0.40	0.014	71.43	0.002	0.045	1.28	0.640	0.3	0.55	3.88	0.30	0.80	0.03	11.00	0.44	1.24	0.638	0.50	3.37

Note: Data for the channel: y, n & S; data for the silt basin and forebay: V_h, V_d, d

Table 7.3 Distribution of steel reinforcement in the silt basin and forebay tank

Q	d_d	W = L_e=L_s	d_1	L_d	d_2	D_i	d	d_3	E	M	On the floor Longitudinal – cross-section of rods[1] and their spacing	In the walls Cross-sectional and/or vertical wall	Longitudinal and/or special reinforcement
m³/s	m	m	m	m	m	m	m	m	m	m			
0.009	0.15	0.20	0.16	3.00	0.28	0.075	0.15	0.49	0.10	0.10	1/4"∅@ 0.11 m spacing	1/4"∅@ 0.20 m spacing	1/4"∅@ 0.15 m spacing
0.014	0.17	0.28	0.19	3.40	0.33	0.096	0.15	0.58	0.10	0.10	1/4"∅@ 0.15 m	1/4"∅@ 0.22 m	1/4"∅@ 0.18 m
0.022	0.19	0.38	0.22	3.80	0.38	0.117	0.15	0.68	0.10	0.10	1/4"∅@ 0.14 m	1/4"∅@ 0.25 m	1/4"∅@ 0.20 m
0.026	0.20	0.43	0.24	4.00	0.40	0.129	0.15	0.73	0.10	0.10	Alternate 1/4"∅ and 3/8"∅ rods @ 0.16 m spacing	1/4"∅+3/8"∅@ 0.30 m spacing	1/4"∅+3/8"∅@ 0.20 m spacing
0.031	0.21	0.49	0.26	4.20	0.42	0.140	0.15	0.78	0.10	0.10	1/4"∅+3/8"∅@ 0.18 m	1/4"∅+3/8"∅@ 0.30 m	1/4"∅+3/8"∅@ 0.20 m
0.042	0.23	0.61	0.29	4.60	0.47	0.164	0.20	0.94	0.12	0.12	1/4"∅+3/8"∅@ 0.23 m	1/4"∅+3/8"∅@ 0.30 m	1/4"∅+3/8"∅@ 0.23 m
0.056	0.25	0.75	0.32	5.00	0.52	0.189	0.20	1.05	0.12	0.12	3/8"∅@ 0.17 m	3/8"∅@ 0.30 m	3/8"∅@ 0.25 m
0.101	0.30	1.12	0.40	6.00	0.64	0.254	0.20	1.34	0.12	0.12	3/8"∅+1/2"∅@ 0.22 m	3/8"∅+1/2"∅@ 0.30 m spacing	3/8"∅+1/2"∅@ 0.25 m
0.164	0.35	1.57	0.48	7.00	0.76	0.323	0.25	1.71	0.15	0.15	3/8"∅+1/2"∅@ 0.25 m	3/8"∅+1/2"∅@ 0.35 m	3/8"∅+1/2"∅@ 0.30 m
0.248	0.40	2.06	0.56	8.00	0.88	0.397	0.30	2.09	0.18	0.18	1/2"∅@ 0.30 m	1/2"∅@ 0.35 m	1/2"∅@ 0.30 m
0.353	0.45	2.62	0.64	9.00	1.00	0.474	0.35	2.48	0.20	0.20	1/2"∅@ 0.31 m	1/2"∅@ 0.32 m	1/2"∅@ 0.30 m
0.483	0.50	3.22	0.72	10.00	1.12	0.555	0.40	2.90	0.20	0.20	1/2"∅@ 0.30 m	1/2"∅@ 0.30 m	1/2"∅@ 0.30 m+column+beam
0.502	0.53	3.16	0.77	10.60	1.19	0.565	0.50	3.04	0.25	0.25	1/2"∅@ 0.30 m	1/2"∅@ 0.30 m	1/2"∅@ 0.30 m+column+beam
0.640	0.55	3.88	0.80	11.00	1.24	0.638	0.50	3.37	0.25	0.25	1/2"∅@ 0.30 m	1/2"∅@ 0.30 m	1/2"∅@ 0.30 m+column+beam

1/4" = 0.6 cm, 3/8" = 1 cm, ½" = 1.3 cm

Table 7.4 Dimensions for the different components of a forebay tank according to the flow (m^3/s) of the MHS

Qm^3/s	Dimensions of the forebay (m)					Measurements of the roof slab (m)				Steel in the roof slab	
	Inner measurements of the tank					Breadth	L_i	A_{ext}	L_{ext}	In the direction of W	In the direction of L_i
	W=L	d_T	e wall	d_d	B_f						
0.042	0.61	0.41	0.10	0.23	0.10	0.08	0.42	0.81	0.52	1 Ø 3/8"@ 0.15 m	1 Ø 3/8"@ .20 m
0.056	0.75	0.45	0.10	0.25	0.10	0.10	0.54	0.95	0.64	1 Ø 3/8"@ 0.16 m	1 Ø 3/8"@ .22 m
0.101	1.12	0.55	0.12	0.30	0.15	0.10	0.86	1.36	0.98	1 Ø 3/8"@ 0.18 m	1 Ø 3/8"@ .22 m
0.164	1.57	0.62	0.12	0.35	0.15	0.12	1.27	1.81	1.39	1 Ø 3/8"@ 0.20 m	1 Ø 3/8"@ .25 m
0.248	2.06	0.73	0.12	0.40	0.18	0.15	1.70	2.30	1.82	1 Ø 3/8"@ 0.20 m	1 Ø 3/8"@ .26 m
0.353	2.62	0.78	0.15	0.45	0.18	0.15	2.27	2.92	2.42	1 Ø 1/2"@ 0.20 m	1 Ø 3/8"@ .25 m
0.483	3.22	0.88	0.15	0.50	0.20	0.18	2.79	3.52	2.94	1 Ø 1/2"@ 0.20 m	1 Ø 3/8"@ .25 m
0.502	3.16	0.91	0.20	0.53	0.20	0.18	2.79	3.56	2.99	1 Ø 1/2"@ 0.20 m	1 Ø 1/2"@ .25 m
0.640	3.88	0.95	025	0.55	0.22	0.20	3.45	4.38	3.70	1 Ø 1/2"@ 0.20 + edge beam	1 Ø 1/2"@ .25 m

Source: Author

Construction process

The same steps followed for the silt basin were repeated, except for the greater height of the excavation and formwork. The formwork needs more rigidity and bracing and the established safety standards and technical specifications must be applied.

Once the layout, stake out, excavations and formwork have been completed and the steel prepared and installed, the following steps are recommended before or while laying the concrete and/or stripping the forms off the walls and the roof.
- Lay all the concrete in one day so that the entire structure is monolithic.
- Form a ring over the surface of the pipe with binding wire, winding it several times to ensure a better adherence of the concrete to the PVC pipe and thus avoid leaks.
- Place the specified level in the wall formwork to control the total height of the concrete casing.
- Mark the correct position and anchor the bolts to hold down the screen on the wall supporting it.
- Coat the walls and the tapered inlet pipe to the penstock, then the floor, trying to prevent any debris from entering the pipes.
- Cure the walls and floor for at least 10 days.
- Place a wooden panel at the mouth of the pipe to prevent any casual or intentional entry until the protocol or operating tests are carried out.
- Fit in the screen and whatever safety device is required.
- For security purposes, build a fence around the silt basin and forebay tank to prevent children or small animals from entering, because they could suffer serious harm.
- Supervise all the activities involved in the construction of the forebay tank.

Technical specifications

The specifications are the same as for the silt basin in terms of the quality of the concrete, steel and soil, but because the walls are higher and materials such as the PVC pipe, screen, etc. are required, the following must be added:
- When the penstock is embedded in the wall of the forebay tank it must be sealed to prevent leakage between the pipe contact surface and the surrounding concrete.
- Inside the formwork, insert a cone-shaped mould before the mouth of the pipe. The pipe will be inserted into 2/3 of the wall's thickness with the (specified) angle formed with the horizontal (designed in the vertical plan).
- The tapered section entering the pipe will be perfected during the final finishing work.
- During the formwork, make sure no concrete residues enter the penstock.

106 DESIGNING MINI AND MICRO HYDROPOWER SCHEMES

Photograph 7.1 Forebay tank, silt basin and complementary components

- The horizontal direction of the pipe and the wall in which the forebay tank will be embedded must form a 90° angle.
- The formwork of the walls must be rigid and technical recommendations regarding the quality and minimum dimensions of each component (props, wedges, bracing, etc.) must be applied.
- The props for the formwork of walls 3 m high or taller must be done when the floor has been laid, to prevent the formwork from shifting due to the pressure of the fresh concrete (settlement).
- The formwork of the roof must lean on straight props separated by at least 0.80 m and braced sideways, with a bevel of at least 3″ (7.6 cm) × 3″ (7.6 cm) if they are over 3 m tall.
- When the walls are exposed to the thrust of the earth, the forms must be stripped from the walls after a minimum of seven days and from the roof after 10 days, bearing in mind that the temperature of the environment is higher than 10° C.
- The concrete within the forms must be no more than 30 cm high and in 30–45 cm high layers. In addition, the concrete must be compacted between the layers to avoid cracks or porosity.
- The steel coating must be no less than 4 cm on the floor and 2 cm on the roof.
- The coating or inside casing of the walls and floor must have a layer of cement mortar and sand 1:3 and a polished finish of at least 2 cm.

Photograph 7.1 shows the forebay tank and silt basin together, as well as the other complementary components of the Yanacancha Baja MHS in La Encañada, Cajamarca, Peru.

CHAPTER 8
Penstock

The penstock is an enclosed circular section that carries water under pressure all the way from the forebay tank to the turbine. The penstock is located as follows:
- It is situated on the slope, between the forebay tank and the powerhouse. It is visible when steel piping is used and generally invisible in the case of PVC piping which is buried underground to protect it from damage.
- With respect to the relief, advantage must be taken of the curve to benefit from the natural run-off on either side of the piping and the powerhouse in order to spend less on drainage.
- As regards the geology of the soil, rocky, muddy or potentially unstable soil must be avoided for the layout or alignment.

Penstock materials

In micro and mini hydroelectric power schemes, steel pipes, PVC pipes and high density polyethylene (HDPE) pipes are used.

In the majority of cases, high pressure PVC pipes with diameters ranging from 2" (5 cm) to 20" (50 cm) were used, bearing in mind that pipes of different thicknesses and resilience are commercially available. Among the favourable aspects of PVC pipes are their lower cost and anti-corrosive properties, relatively light weight compared with pipes of other materials, and the availability of accessories for changing direction, changing section and coupling to the turbine distributor pipe. In some cases, when the gross head was more than 150 m, steel pipes were also used up to a certain height, as they withstand greater pressure, and then PVC pipes were used higher up.

HDPE pipes have also been used, especially in pico hydroelectric power schemes. The main advantages of this type of pipe are its flexibility and light weight compared with other types, but the pipes must be joined together by thermofusion, for which special equipment is needed and specific accessories are required to connect the pipes to the turbine. The equipment, accessories and their application in isolated areas increase the installation costs.

In Peru, ITDG initially used RU PVC pressure pipes with rigid unions manufactured in accordance with national technical standards established by the Institute for Industrial Technological Research and Technical Standards (ITINTEC), whereby the pipe end and the socket are joined together with PVC glue. To create good joints, it is important to follow the manufacturer's instructions carefully and those provided in the following sections. This will ensure that there are no leakages and there is good resistance to breaks when there is a sudden change of pressure.

For the past few years, industry has been supplying FU PVC (pipes with flexible unions) pressure pipes that do not require glue to joint. Instead the socket has an incorporated rubber ring with a steel core and the pipe end fits in directly, so only a layer of lubricant is applied. Experience has shown that the unions obtained are hermetic and guaranteed to be leak-proof.

It is also worth mentioning that manufacturers can provide FU PVC accessories in accordance with the change of direction of the penstock and the connection to the turbine. These sorts of accessory are ordered once the penstock design is completed and accurate data for changes of direction are available.

PVC has replaced steel pipes for pressure heads lower than 150 m, as the lower installation and maintenance costs are an advantage. In addition, its installation is simpler since no welding equipment or specialized manpower are required. Furthermore PVC pipes need neither support structures nor accessories to absorb expansion stress.

Penstock accessories (Figure 8.1)

The following items are required to join the pressure pipes:
- Curves, depending on the angle determined by the change of direction on the vertical plan or the change of alignment on the plan of the slope (topography of the ground).
- T-pipe with reduction for the vent pipe.

For the connection to the turbine:
- A reduction to gradually change the diameter of the piping to the size of the diameter of the turbine distribution pipe.

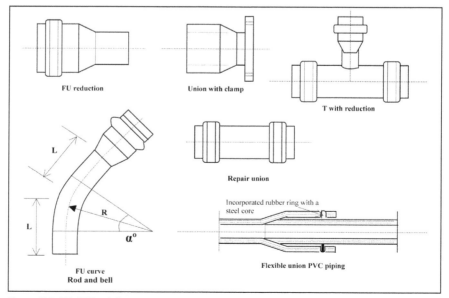

Figure 8.1 FU PVC piping and accessories, standard ISO 4422

- A union with a removable clamp to be attached to the clamp of the turbine distributor pipe.

For maintenance and/or repairs:
- Repair union.

Penstock elements

For steel pipes (Figure 8.2)

- Pipes: they can be made of rolled and electro-welded sheeting or of a single seamless piece.
- Clamps: they are welded onto each end, standardized for each thickness and diameter. They make it easier to join pipes with a rubber washer, screws, rings and bolts.
- Elbow pipes with clamps at both ends.
- Expansion joints: they usually go at the end of each straight section before the penstock changes direction and before the anchor immediately above (see Figure 8.2). They absorb the expansion or contraction of the piping caused by the changes of temperature.
- Valves: for the passage of water. There are different types, but the ones used most often are gate valves.
- Supports: concrete blocks that the steel pipes lean on to spread out the weight and prevent bending.
- Anchors: concrete blocks that fix the piping to the ground to prevent any lateral or vertical shifting.
- Expansion coupling: a metal union to join the pipe to the turbine distribution pipe.
- Vent: to eliminate the air trapped in the penstock before it starts operating and to help dissipate any additional transitory pressure.

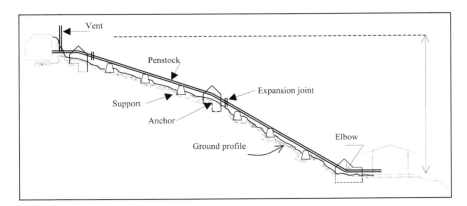

Figure 8.2 Steel pressure penstock

112 DESIGNING MINI AND MICRO HYDROPOWER SCHEMES

For PVC penstocks (Figure 8.3)

- 5 m long pipes for rigid unions, 6 m long pipes for flexible unions (the length may change depending on the national standards within the country of installation).
- Rubber ring with a steel core incorporated or to be fitted in each pipe (for pipes with flexible unions).
- Curves for rigid union or flexible union pipes.
- Concrete anchors.

The entire penstock rests on the channel bed (there is no need for supports for each pipe as is the case for steel pipes). PVC pipes are buried to protect them from ultraviolet rays, changes in temperature and physical damage.

For HDPE pipes, with 2" to 4" diameters (Figure 8.4)

- They are sold commercially as rolled pipes of variable lengths.

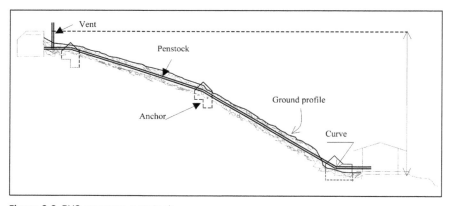

Figure 8.3 PVC pressure penstock

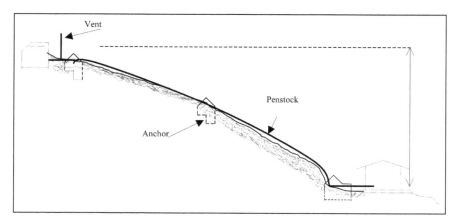

Figure 8.4 HDPE pressure penstock

- Metal accessories such as adaptors, reducers, etc.
- Anchors.

They can be installed on the surface or underground, providing they are aligned and with no undulations. They respond well to ultraviolet rays and changes in temperature and are strong enough to withstand impacts.

Penstock design (see Annexe Examples 9 and 10)

To design the penstock, it is necessary to know the design data, the information processed in the office, the formulas to be employed and the measuring units to be used. Catalogues must be obtained from manufacturers of pipes and accessories to identify the physical or mechanical characteristics, such as the Young's modulus, maximum yield stress (breakage), thickness, diameters and others. Then the respective calculations can be made, making sure that the diameter and thickness of the selected pipes are appropriate to withstand the design pressure and stresses.

The supports and anchors must also be calculated for steel pipes and the anchors for PVC pipes.

Field data and/or office data

The ground profile and the final profile of the penstock must be available, depending on the material to be used, so that the following can be noted.
- H_B = gross head (m); this is the difference in elevation between the surface of the water in the forebay tank and the centre of the jet in the case of action turbines (Pelton, Michell Banki or others). In the case of reaction turbines such as Kaplan or Francis turbines, it is the difference in elevation between the surface of the water in the forebay tank and the surface of the water in the discharge channel.
- H_i = pressure head (m) carried by each section of the penstock.
- L_T = total length of the penstock (m)
- L_i = length of each section of the penstock (m) between anchors, i.e. where the pipe changes direction.
- The angles formed by the sections when the penstock changes direction.
- The level of the turbine shaft, depending on the turbine to be installed (consult the turbine manufacturer or supplier).
- Q = design flow (m³/s) that should reach the turbine.

It is also important to identify the type and characteristics of the soil on the slope in which the turbine will be installed, the anchors and the supports (for steel pipes).

Calculating the inner diameter of the penstock

The calculation of the inner diameter is generally made by a process of trial and error until the right values are found, which ensures the cost effectiveness of the pipes used. This process is necessary because at the outset there are several unknown parameters such as losses and speed. The formula used for calculating the diameter is as follows:

$$d_i = \sqrt{\frac{4 \times Q}{\pi \times V}}$$

To begin the calculations an appropriate value for the water velocity in the pipe is assumed to be between 2 and 3 m/s. Calculate d_i, and, depending on the material selected, make a note in the corresponding catalogue of the nominal or outer diameter and thickness of the pipe wall in order to determine the (tentative) inner diameter for each section of the penstock profile.

Calculating the head loss h_p, (in each section of the pipe)

$$h_p = h_f + h_t$$

where:
h_f = head loss due to friction in the interior wall of the pipe (m).
h_t = head loss due to turbulence when the water enters the pipe, at section changes, in curves, valves, etc.

$$h_f = 0.08 \times \frac{f \times L_i \times Q^2}{d_i^5}$$

f = friction factor (dimensionless)
L_i = length of pipe section (m)
Q = design flow (m³/s)
d_i = inner diameter of the pipe (m)

Calculating f

Moody's diagram is used for loss of friction in steel or PVC pipes. To use the Moody diagram, the following values must be determined first:

$$1.27 \times \frac{Q}{d} \quad \text{and} \quad K/d_i$$

where:
K = surface roughness of the inner walls of the pipe (m) (Table 8.1; note that the values here are in mm)
d_i = inner diameter (m). To identify the ratio, both must be expressed in the 'same unit'
With these values, read the value of f in the Moody diagram and determine the value of h_f, replacing those values

PENSTOCK

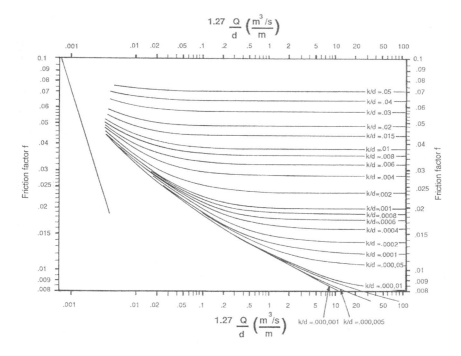

Figure 8.5 Moody diagram to find the friction factor on the inner wall of the penstock

Table 8.1 Surface roughness values '*K*' (mm)

Penstock material	State and age of the pipes		
	Good < 5 years	Normal 5–15 years	Poor < 15 years
PVC, polyethylene, polyester, resin with fibreglass	0.003	0.01	0.05
Concrete	0.06	0.15	1.5
Smooth steel:			
a) Unpainted	0.01	0.1	0.5
b) Galvanized	0.06	0.15	0.3
Wrought iron			
a) New:	0.15	0.3	0.6
b) Old:			
b.1 Slight corrosion	0.6	1.5	3.0
b.2 Moderate corrosion	1.5	3.0	6.0
b.3 Severe corrosion	6	10	20

Source: Harvey, 1993

If the Moody diagram is not available, use the following formula:

$$f = \frac{0.25}{\left[\text{Log}\left(\dfrac{\text{Relative roughness}}{3.7} + \dfrac{5.74}{(\text{Reynolds No.})^{0.9}}\right)\right]^2}$$

To replace the expressions equivalent to the relative roughness and the Reynolds number, the formula is expressed as follows:

$$f = \frac{0.25}{\left[\text{Log}\left(\dfrac{\frac{k}{d_i}}{3.7} + \dfrac{5.74}{\left(\frac{V \times d_i}{\mu}\right)^{0.9}}\right)\right]^2}$$

where:
K = surface roughness depending on the material (mm)
d_i = inner diameter of the pipe in the same (mm)
V = velocity of the water in the section (m/s)
$¼$ = kinematic viscosity of the water (m²/s)

Calculating the head loss due to turbulence

$$h_t = \frac{V^2}{Vg}(K_1 + K_2 + \ldots K_n)$$

where:
V = velocity of the water in the penstock in the section under consideration
$K_i = K_1, K_2, \ldots\ldots K_n$ is a factor associated with elbows, curves, valves, changes of direction, etc.
g, is the gravity acceleration = 9.81 m/s²

This step can be avoided if the calculated values are much lower than the friction loss values.

Once those values are determined, find $h_p = h_f + h_t$

Calculating the head friction losses, in percentage terms

$$\text{Losses} = \frac{h_p}{H_B} \times 100$$

The percentage of losses in the penstock affects the output power of the scheme: the higher the losses the lower the power output. To decrease the losses we need to increase the size of the diameter of the penstock, but that increases the cost of the penstock. There are two main criteria used to decide the percentage losses in

PENSTOCK

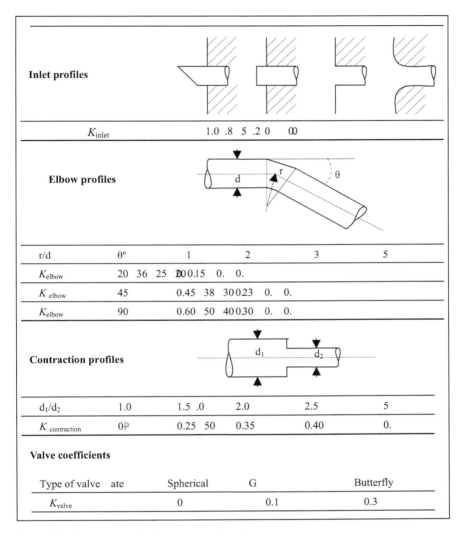

Figure 8.6 Head loss coefficients due to turbulence
Source: Harvey, 1993

the penstock: (i) to have 'small losses' in order to get the maximum power output possible when the available resources (head and flow) are limited; and (ii) to have the most cost-effective size when there are plenty of available resources (head and flow). Establishing the most cost-effective size requires several calculations. From experience, depending on the quality of the materials, the length of the pipe, pressures, and other factors, the cost-effective sizes are generally found when the total friction losses in the penstock are below 10 per cent.

Calculating the thickness of the penstock wall in each section

The wall thickness should be designed not only for normal running of the system but also to respond to the following.

Surge pressures: Δh

These are transitory or temporary pressures that occur suddenly when the flow stops. In hydroelectric schemes this can occur when valves are closed suddenly or there is a blockage in the penstock.

Thinning of the pipe walls

Thinning of the inner walls of the pipes while in service, caused by corrosion, wear and tear or manufacturing defects.

The transitory pressure in m, is calculated by the following formula

$$\Delta h = \frac{a \times V}{g}$$

where:
V = velocity of the water in the section of the pipe (m/s)
a = pressure wave propagation speed (m/s)
g = gravity acceleration = 9.81 (m/s^2)

Pressure wave propagation speed a, depends on the diameter, thickness and material of the pipe. For a referential approximation, the following values can be employed:
Soft steel piping a = 900 m/s
Wrought iron piping a = 1,250 m/s
PVC   penstock a = 350 m/s

For more accurate calculations, use the following formula

$$a = \frac{1420}{\sqrt{1 + \left(\dfrac{E_{ag} \times d_i}{E_{pvc} \times t}\right)}}$$

where:
E_{ag} = bulk modulus of water (kg/cm^2)
E_{pvc} = Young's modulus of the PVC (kg/cm^2)
d_i = inner diameter (m)
t = thickness of the pipe (m)

If the pipes are made of steel or another material, use the corresponding Young's modulus, in accordance with the table of values provided by the manufacturer's (catalogues).

Thickness T

This value is the actual thickness of the penstock wall when it is new. Its calculation is made using the following equation; however after some years of operation the thickness value may be reduced owing to external factors such as corrosion or other environmental effects. Therefore for the purpose of the calculations the parameter 't' is introduced as the effective thickness of the penstock; see its equation below.

$$T = \frac{5 \times f_s \times h_T \times d_i \times K_j}{\sigma} + K_c$$

where
T = expected thickness (mm)
f_s = safety factor depending on the piping material
f_s = 3.5 is usually recommended
f_s = 3 is recommended for PVC material
If the cost of the piping is high, minimum = 2.5

This decision applies when the pipes are short, the gross head is low, the slope is slight, the valve operation is slow, the staff are trained and the piping is installed by qualified staff.

$h_T = H_i + \Delta h$

h_T = 'maximum gross head' (m), in the considered section
H_i = gross head or pressure in the section (m).
Δh = surge pressure (m), in the considered section.
d_i = inner diameter of the penstock (m).
σ = maximum yield stress of the considered material (N/m²)
K_j = thickness correction factor, depending on the piping material

For steel pipes
K_j = 1.1 mm, welded union factor
K_j = 1.2 mm, rolled and welded sheeting factor
K_c = 1.0 mm, corrosion factor (1 mm every 10 years)

For PVC pipes
K_j = 1.0 mm
K_c = 1 mm every 10 years, wear and tear factor

Safety factor (f_s)

As in other engineering structures, it is convenient to use a safety factor. The following steps are recommended for the introduction of the safety factor.

Estimate the value of the effective thickness (t) using the formula for T with simplifications and logical deductions

$$T = \frac{5 \times f_s \times h_T \times d_i \times K_j}{\sigma} + K_c$$

For PVC: $K_j = 1$
Then: $T - K_c = t$

Therefore the equivalent formula for PVC is:

$$t = \frac{5 \times f_s \times h_T \times d_i}{\sigma}$$

Analysing this formula, the safety factor must be determined

$$f_S = \frac{t \times \sigma}{5 \times h_T \times d_i}$$

on condition that in each section in which it is applied, $f_s = \geq 3$; and the units to be used are:

t, in mm
σ, in N/m²
h_T, in m
d_i, in m

For steel: the units are the same.
Note: In both cases, if the $\tilde{A}$ units are expressed in kg-f/cm², the equivalent formula is:

For PVC: $$f_S = \frac{t \times \sigma}{5 \times h_T \times d_i \times 10}$$

For steel: $$f_S = \frac{t \times \sigma}{5 \times h_T \times d_i \times K_j \times 10}$$

Data taken from the tables and catalogues of manufacturers of steel and PVC pipes

- E_{AG} = bulk modulus of water = 21,000 kg-f/cm²
- μ = kinematic water viscosity (15°C) = 1.14 x 10⁻⁶ kg-f/cm²
- E_{AC} = steel Young's modulus = 2,000,000 kg-f/cm²
- E_{PVC} = PVC Young's modulus = 28,000 kg-f/cm²
- σ_{AC} = maximum yield stress of the steel (breakage) = 3500 kg-f/cm²
- σ_{PVC} = maximum yield stress of the PVC (breakage) = 560 kg-f/cm²
- K_{AC} = surface roughness of the steel = 0.1 mm
- K_{PVC} = surface roughness of the PVC = 0.01 mm

- d_e = nominal outer diameter of the pipe (m)
- d_i = inner diameter of the pipe (m)
- t = effective thickness of the pipe (mm)

The physical and mechanical characteristics and dimensions of the different kinds of pressure pipe commercially available are shown in Tables 8.5–8.8 at the end of this chapter.

Supports (see Annexe Example 11)

The supports are usually coarse concrete blocks that support the weight of the water and the steel penstock.

They are placed at uniform distances on every section of the piping. A distance between supports of halfway along the pipe is recommended to prevent bending, unless the project designer specifies another distance. Pipes joined by clamps are considered a single pipe.

As steel pipes are not buried, the temperature will cause them to contract or expand, therefore the supports must allow this lengthwise movement, bearing in mind that this creates friction between the steel and the concrete on the contact surface and, as the support transmits its own weight as well as that of the pipes and the ground water, it should have the capacity to prevent any settlement.

Design of the supports

First, select a support profile with its respective dimensions, depending on the topography of the soil.

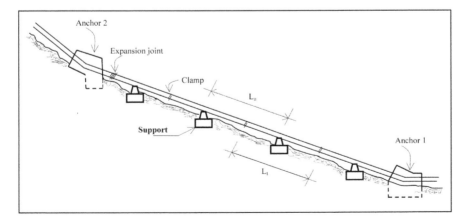

Figure 8.7 Location of supports between anchors, in a section of the penstock

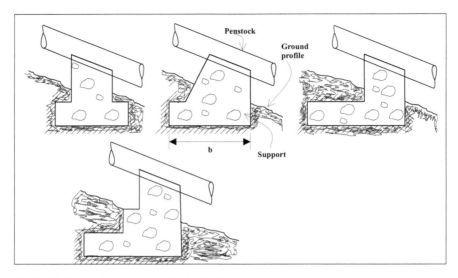

Figure 8.8 Types of support profile for steel pipes

Secondly, determine the stresses involved. Two cases are presented: When the pipes expand, the force of the friction between the material of the pipe and the block is upwards, in the direction of the pipe; when it contracts, *f* changes direction downwards, in the direction of the pipe.

Total weight of the pipe and the water act on the contact surface between the pipe and the block (W_{ta}, kg-f or N)

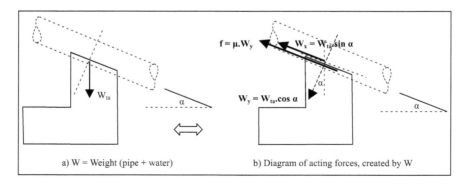

a) W = Weight (pipe + water)   b) Diagram of acting forces, created by W

Figure 8.9 Diagram of acting forces when the pipes expand

Forces that act on the concrete block

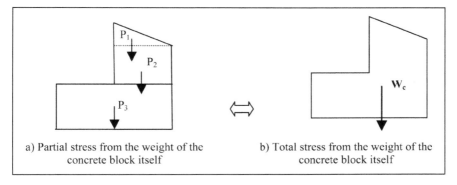

Figure 8.10 Diagram of the forces caused by the actual weight of the concrete block

Forces involved for calculation purposes

$f = \mu . W_y$
$W_y = W.\cos \alpha$
W_c = weight of the concrete

$W_{x'}$ is not required in the calculation of the support, but it is for the anchor.

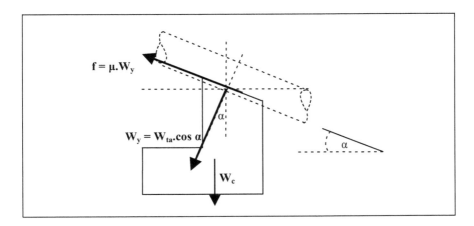

Figure 8.11 Forces involved in the calculation of the support

Stability of the support

1) Bending

   Condition: $\dfrac{\Sigma \text{ stablizing } M}{\Sigma \text{ bending } M} \geq 2$

   Σ stabilizing moments = vertical stress moments
   Σ bending moments = horizontal stress moments
   The moments are applied at the same point.

2) Sliding

   Condition: $\Sigma F_x < \mu_T \cdot \Sigma F_y$

   μ_T = soil and concrete friction coefficient, depending on the type of soil (Table 2.2, varies between 0.45 and 0.70)

3) Stability of the foundation soil

   Condition: $\sigma_T > \sigma_{max} > \sigma_{min}$

   a) $\sigma_{max} = \dfrac{R_y}{A}\left[1 + \dfrac{6e}{b}\right]$

   b) $\sigma_{min} = \dfrac{R_y}{A}\left[1 - \dfrac{6e}{b}\right]$

   σ_{max} = maximum bearing capacity transmitted to the foundation soil (kg-f/cm^2)
   σ_{min} = minimum bearing capacity transmitted to the foundation soil (kg-f/cm^2)
   σ_T = admissible bearing capacity of the foundation soil (kg/f/cm^2)
   R_y = force of the soil reaction to the stress and load it bears (kg-f)
   A = area of the surface of the base of the block (cm^2)
   e = eccentricity of the vertical reaction (m)
   $e = X - b/2$
   X = distance from moment origins to the point of action Ry on the base of the block
   b = length of the base of the block profile (m)

Complementary definitions and formulas

- $W = W_t + W_a$
  W = weight of the pipe and the water per linear metre (kg-f /m or N/m)
- W_t = weight of the pipe, per linear metre (kg-f /m or N/m).
  Weight of the pipe = specific weight of the steel × volume

$$W_t = \gamma_t \times \dfrac{\pi}{4}\left[d_e^2 - d_i^2\right]$$

γ_t = specific weight of the pipe material (kg-f/m^3 or N/m^3)
γ_{steel} = 7,860 kg/m^3
d_e = outer diameter of the pipe (m)
d_i = inner diameter of the pipe (m)

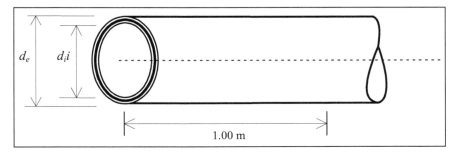

Figure 8.12 Section of a penstock showing internal and external diameter

- W_a = weight of the water inside the pipe, per lengthwise metre
  Weight of the water inside the pipe = specific weight of the water × volume

$$W_a = \gamma_a \times \frac{\pi \times d_i^2}{4}$$

γ_a = specific weight of the water = 1,000 kg-f/m³.

- The weight of the pipe and the water to be supported by the support is $W_{ta} = W \times L_t$
- W_x = horizontal component of W_{ta} (kg-f or N). It runs parallel to the pipe
- $W_x = W_{ta} \cdot \sin \alpha$
- α = angle of inclination of the pipe from the horizontal
- W_y = vertical component of W_{ta} (K-f or N). At right angles with the pipe
- $W_y = W_{ta} \cdot \cos \alpha$
- f = force of friction between the concrete and the steel pipe (K-f or N)
- $f = \mu \cdot W_y \rightarrow \mu$ = friction coefficient. For the steel and concrete, $\mu = 0.5$–0.6
- L_t = length of the pipe = L_a = distance between the central lines of the supports (m)

Recommended value: $L_a \leq 8$ m. There should be no bending or sagging.

The admissible sag for the girders is:

$$\delta_{admissible} = \frac{L_a}{360}$$

Maximum sag δ, is calculated with the formula:

$$\delta = \frac{5 \times W \times L_a^4}{384 \times E \times I}$$

where
δ = sag (m)
W = load, distributed by linear metres = (weight of the pipe and water)/m
L_a = length of the pipe in m

E = young's modulus of the piping material
$E_{steel} = 200 \times 10^9 \text{ N/m}^2 = 20 \times 10^9 \text{ kg-f/m}^2$
I = inertia moment of the section, in m^4

$$I = \frac{\pi}{64}\left(d_e^4 - d_i^4\right)$$

- W_c = weight of the concrete, in kg-f, or N.
  Weight of the concrete = Specific weight × Volume
  γ_c = 2,200 kg-f/m^3 (of the coarse concrete)
  = 2,400 kg-f/m^3 (of the reinforced concrete)

Anchors (see Annexe Example 12)

The anchors are coarse concrete or reinforced concrete blocks that anchor the penstock to the ground. They do not allow the pipes to move vertically or sideways, whether they are steel or PVC pipes.

They are usually located at both ends of each section of the pipe, where the pipe changes direction. They are also located within the same stretch when a change of section was determined in the piping design.

In view of the change of direction and the movement of the water inside the pipe, there are two types of anchor, one facing out and the other facing in.

The anchor faces out when the pipe appears to break away from the ground when it changes direction. It faces in when the pipe tries to continue towards the ground when changing direction, as though it was nailed down (Figure 8.14).

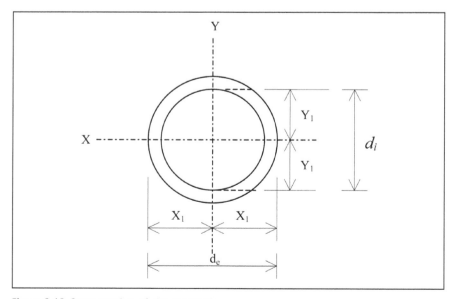

Figure 8.13 Cross section of the penstock

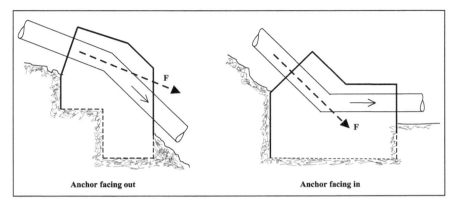

Figure 8.14 Types of anchor for different landscape profiles

Anchor designs

- Select the type of anchor.
- Determine the corresponding measurements.
- Make a note of the following on the pipe profile, for the anchor considered:
  - Angle α, formed by the pipe with the horizontal upstream from the anchor, then angle β formed by the pipe and the horizontal downstream from the anchor.
  - Pressure head H_i
  - L_i of the pipe for every intervening force
- Determine each of the intervening forces
- Verify the stability to bending, sliding and settlement of the soil, for both cases. When the pipe is subject to expansion or contraction, the calculation is similar to the calculation of the supports.

Forces intervening in the anchors

F_1 and F_2 are components of the total weight produced by the penstock and the water contained in it (see direction of forces and formulas).

F_1 = pipe weight + water weight

$F_1 = W \times L_1 \times \cos \alpha$

W = in kg-f/m = (weight of pipe + weight of water) × 1 m long
L_1 is half the length of pipe section $L_i = L_i/2$, in m

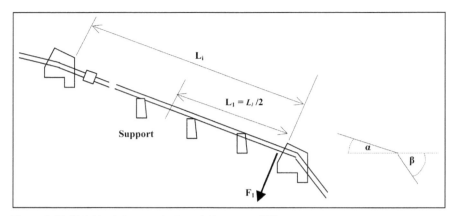

Figure 8.15 Weight of the penstock and the water (F1)

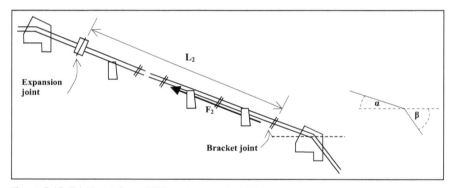

Figure 8.16 Frictional force (F2) and expansion joint

F_2 = force of friction between the pipe and the supports. For this force to be relived, there should be at least one upward expansion joint

$F_2 = \mu \times W \times L_2 \times \cos \alpha$

μ = friction coefficient between the concrete and the steel = 0.5
W = in kg-f/m. = (weight of pipe + weight of water) x 1 m long
L_2 = length between the first bracket joint near the anchor and the expansion joint, near the top anchor. In practical terms, it is the length of the section minus the length of the pipe, in m.

F_3 = due to the hydrostatic pressure on the elbow pipe or on the curve when it changes direction

$F_3 = 1.6 \times 10^3 \times H_i \times d_i^2 \times \sin \dfrac{\beta - \alpha}{2}$

H_i = pressure head in the anchor, in m
d_i = inner diameter of the pipe
F_3, in kg-f

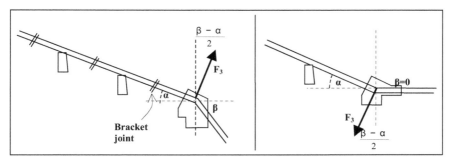

Figure 8.17 Force due to the hydraulic pressure on the anchor (F3)

F_4 = penstock weight component, parallel to the pipe

$F_4 = W_t L_4 \times \sin \alpha$

F_4, in kg-f
W_t = weight of the pipe in kg-f/m
L_4, in m (Practically the total length of the section, minus half the length of one pipe)

F_5 = stress caused by changes of temperature in the piping.
This force is created when the piping is on the surface and no expansion joint has been fitted

$F_5 = 31 \times d_i \times t \times E \times a \times \Delta T$

F_5, in kg-f;
d_i, in m
t, in mm (thickness of the pipe)
E, Young's modulus of the pipe material, in kg-f/cm^2
a, linear expansion coefficient of the piping, in °C^{-1}

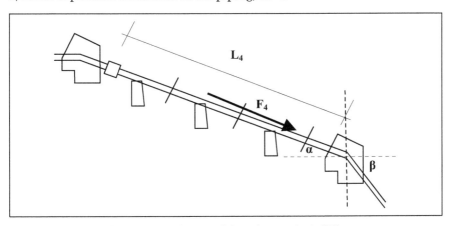

Figure 8.18 Weight of penstock acting parallel to the penstock (F4)

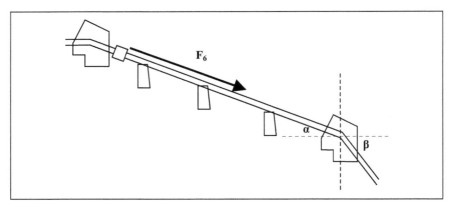

Figure 8.19 Frictional forces in the expansion joint (F6)

This force is = 0 when the section of steel piping has an expansion joint
PVC piping is buried. It is not exposed to temperature changes, therefore $F_5 = 0$

F_6 = Force of friction in the expansion joint
This occurs between the washer and the parts of the expansion joint. It is a two-way force – when the piping expands and when it contracts.

$F_6 = 3.1 \times d_i \times C$

d_i = inner diameter of the piping, in m
C = Friction in the expansion joint, expressed in kg-f/unit of the circumference
Approximately:
$F_6 = 10 \times d_i$
Note: d_i, in mm
F_6, in kg-f

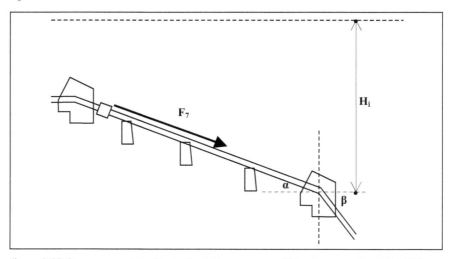

Figure 8.20 Stress caused by the hydrostatic pressure within the expansion joint (F7)

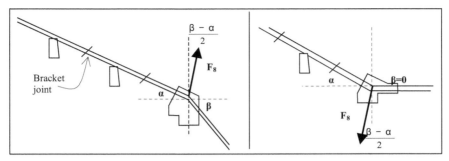

Figure 8.21 Stress caused by the change in direction of the water moving in the elbow pipe or bend

F_7 = Stress caused by the hydrostatic pressure within the expansion joint. It tries to separate the piping into the two sections connected by the joint.

$F_7 = 3.1 \times H_i \times d_i \times T_t$

H_i = pressure head in the piping at the level of the anchor, in m
d_i = inner diameter of the piping, in m
T = thickness of the pipe, in mm

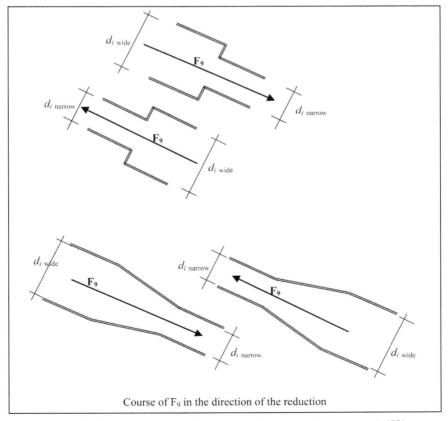

Figure 8.22 Hydrostatic forces due to the change of section of the penstock (F9)

DESIGNING MINI AND MICRO HYDROPOWER SCHEMES

Table 8.2 Summary of the forces intervening in the calculation of the anchor (see Figure 8.23)

F	Definition	Formula	L_i or H_i	Direction/course	Units
F_1	Component of W	$F_1 = W.L_1.\cos\alpha$	$L_1 = L_i/2$	At a right angle with the pipe/downwards	$W = (W_t + W_a)$, in kg-f/m F_1, in kg-f L_1, in m
F_2	Friction between the pipe and supports	$F_2 = \mu.W.L_2.\cos\alpha$	$L_2 = L_i - L_t$	Parallel to the pipe Expansion: downwards Contraction: upwards	W, in kg-f/m L_2, in m $\mu = 0.5$ friction coeff. between the steel and concrete L_t = Length of the pipe (m)
F_3	Hydrostatic pressure when changing direction	$F_3 = 1.6.10^3 . H_i . d_i^2 . \sin[(\beta-\alpha)/2]$	H_i = gross head in the anchor	According to angle $(\beta-\alpha)/2$ with the vertical, downwards	d_i = inner diam. of the pipe H_i and d_i, in m F_3, in kg-f
F_4	Component of W_t	$F_4 = W_t.L_4.\sin\alpha$	$L_4 = L_i - L_t/2$	Parallel to the pipe/downwards	F_4, in kg-f W_t, in kg-f/m L_4, in m
F_5	Due to changes in temperature in the piping	When an expansion joint was not considered for the section $F_5 = 31 . d_i . t . E . a' . \Delta T$ $\Delta T = T_1 - T_2$ $(F_5 = 0)$, **If an expansion joint was considered for the section**	- - - - - - -	Parallel to the pipe Expansion: downwards Contraction: upwards	F_5, in kg-f; d_i, in m t, in mm (thickness of the pipe) E, Young's modulus of the pipe material, in kg-f/cm² a' = linear expansion coefficient of the pipe, in °C⁻¹

PENSTOCK   133

F	Definition	Formula	L_i or H_i	Direction/course	Units
F_6	Force of friction in the expansion joint	$F_6 = 3.1 d_i \cdot C$ Approximately: $F6 = 10 \cdot d_i$ d_i in mm	- - - - - - - -	Parallel to the pipe Expansion: downwards Contraction: upwards	d_i = inner diameter of the pipe, in m C = friction of the expansion joint, expressed in kg-f/unit length of circumference F_6, in kg-f
F_7	Hydrostatic pressure in the expansion joint	$F_7 = 3.1 \cdot H_i \cdot d_i \cdot t$	H_i = gross head in the anchor	Parallel to the pipe downwards	d_i = inner diameter of the pipe, in m t = thickness of the pipe, in mm
F_8	Change in direction of the quantity of moving water	$F_8 = 250(Q/d_i)^2 \cdot \sin[(\beta-\alpha)/2]$ Small value. Can be averted.	- - - - - - - -	Same as F_3	Q in m³/s d_i = inner diameter, in m
F_9	Change in the diameter of the pipe from wide to narrow, with a reducing device	$F_9 = 1 \times 10^3 \cdot H_i \cdot (di^2_{wide} - di^2_{narrow})$ Usually insignificant. Can be averted.	H_i = Pressure head in the change of section.	Parallel to the pipe Downwards: diameter changes from wide to narrow Upwards: diameter changes from narrow to wide	H_i in m $d_{i\,wide}$, in m $d_{i\,narrow}$, in m

F_8 = Stress caused by the change in direction of the quantity of water moving in the elbow pipe or bend. The direction is the same as F_3.

$F_8 = 250(Q/d_i)^2 \times \sin[(\beta - \alpha)/2]$

This is a minor force compared with the others, therefore it can be averted.

F_9 = Stress caused by the change in the inner diameter, due to diameter reduction in the penstock. It acts in the direction of the reduction, parallel to the piping.

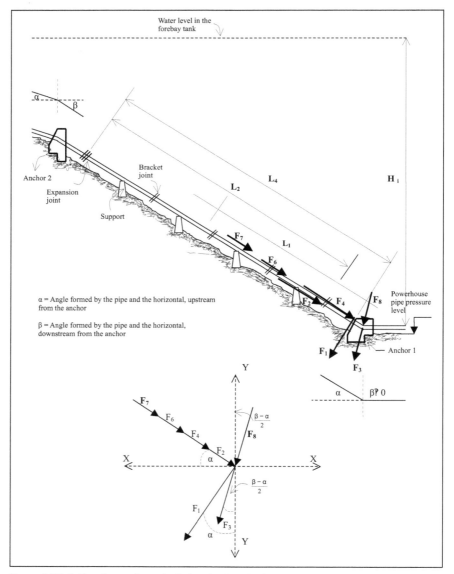

Figure 8.23 Acting forces for calculating Anchor 1, Case I, when the pipe expands

$F_9 = 1 \times 10^3 \times H_i \times (d_{i\,\text{wide}}^2 - d_{i\,\text{narrow}}^2)$

H_i = Pressure head on the reduction, in m
$d_{i\,\text{wide}}$ = diameter of the wider pipe, in m
$d_{i\,\text{narrow}}$ = diameter of the narrower pipe, in m

Construction process

PVC pipes

The following steps are guidelines for a typical site.
- At the worksite, the pipes must be stored under a roof and on a flat surface, to prevent them being deformed by infrared radiation.
- Remove brush and clear the strip of land for the penstock.
- Draw and stake out the centre line of the piping, making sure that the pipe and the adjacent wall of the powerhouse and the forebay tank form a right angle.
- Mark the width of the ditch, taking the centre line as a reference. The minimum width recommended is the length of the pipe diameter plus 60 centimetres.
- Along the centre line, stake out and fix key and referential levels, in accordance with the penstock profile, using a topographic level or inclinometer and a measuring tape. Then draw the ground profile, key levels and the penstock profile to scale, in accordance with the project plan.
- In the powerhouse, determine the inlet level of the pipe with respect to the level of the finished floor. In the forebay tank, determine the

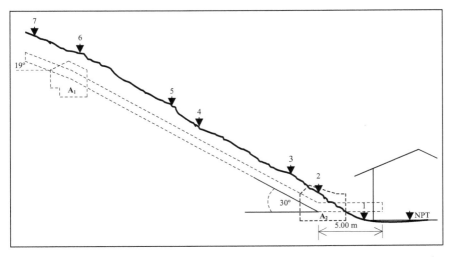

Figure 8.24 Positioning of stakes and levels in the centre line of the piping drawn on the ground

136 DESIGNING MINI AND MICRO HYDROPOWER SCHEMES

Photograph 8.1 Levelling and fitting the penstock at the entrance of the powerhouse, with respect to the level of the finished floor and the location of the turbine

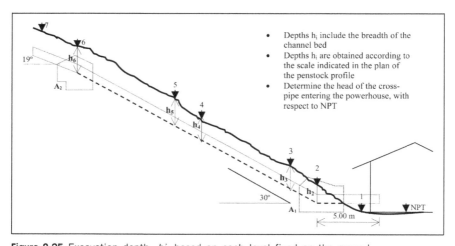

Figure 8.25 Excavation depth, hi, based on each level fixed on the ground

embedding level of the pipe with respect to the level of the water surface and the bottom floor.
- Dig the ditch for the piping to the depth indicated in the plan of the ground and piping profile. From this plan and, bearing in mind its scale, obtain depths h_i, to be measured at each key level.
- Shape and align the excavated ditch.

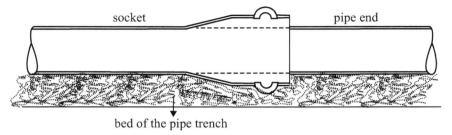

Figure 8.26 Penstock joints

- Determine the flush levels of the channel bed and the changes of direction. A trench bed is necessary when the soil is rocky and the surface irregular; to obtain the required angle at the bottom of the ditch, display the curve of the corresponding anchor.
- Build the trench bed with the specified compacted material, trying to ensure that the union between the pipe end and the socket allow the pipes to rest on it in a uniform manner along its entire length (Figure 8.26).
- When installing the piping and accessories, discard those that are flawed or with manufacturing defects. Begin in the powerhouse and work towards the forebay tank, with the socket tube facing up hill to make it easier to join the pipe end.

PVC pipes can be joined two by two and then in the ditch, for diameters of up to 8" (20 cm). For larger diameters, it is recommended that they are joined in the ditch itself, one by one. In both cases, resort to the manufacturer's catalogues and recommendations regarding the length of the pipe end to be inserted in the socket.

Rieber union PVC pipes have a rubber ring with a steel core incorporated in the socket (see Table 8.3)

Table 8.3 Socket joint insertion lengths

Nominal diameter (D)		Length of socket	Length of insertion
(mm)	(inches)	L (mm)	l ± 5 (mm)
63	2	120	95
75	2½	130	105
90	3	130	105
110	4	140	115
140	5½	150	125
160	6	170	145
200	8	180	155
250	10	211	185
315	12	250	225
355	14	250	225
400	16	260	235

138 DESIGNING MINI AND MICRO HYDROPOWER SCHEMES

When reaching the point where the pipe changes direction, only one part of the pipe is needed to connect to the curve; shear the pipe and bevel the sheared end before inserting it in the pipe. The socket part of this pipe must face up hill so that the bent pipe end can be inserted.

When joining these pipes, the outside surface of the pipe end and the inside of the socket must be clean at all times, so that a layer of lubricant can be applied to ease the insertion of the pipe end.

In the powerhouse, ensure the horizontal direction of the pipe to be connected to the turbine distributor pipe (coordinate with the turbine manufacturer).

In the forebay tank, the pipe must be embedded in half of the wall so that the final finish of the inside wall is like a funnel, thus ensuring a minimum head loss previously established in the penstock calculation.

- Install the T-pipe with a reducer in the indicated place, to connect the vent pipe before the forebay tank (see Figure 8.27).
- Align and fit the pipes before putting the anchors in place.
- Build the anchors after determining the layout, stake out, excavation and location of the foundation level and verifying the soil quality (bearing capacity). Fix the curve so that it is not shifted by the pressure of the fresh concrete being poured (see Figure 8.28). Form in accordance with the selected profile shape, preparing the specified proportions of materials and pouring the concrete (it should be left to set for at least 15 days).

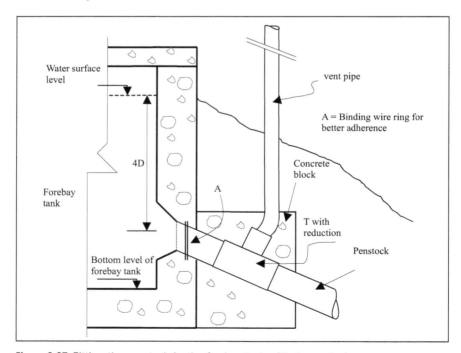

Figure 8.27 Fitting the penstock in the forebay tank with the vent pipe

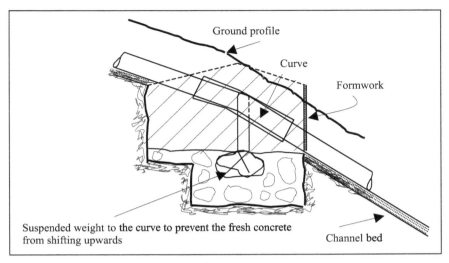

Figure 8.28 Fixing of the curve while laying the concrete

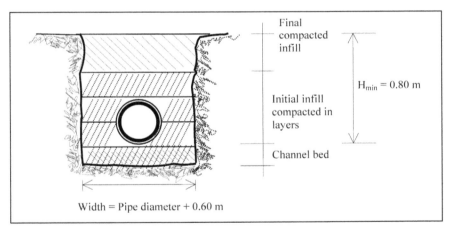

Figure 8.29 Dimensions and covering of the ditch

- Cover the ditch with layers of sifted and compacted material at least 15 cm thick. Start with the sides of the pipe, without altering their alignment, then on top of the pipe (see Figure 8.29).
- Throughout the pipe installation process, the supervisor and work force must ensure that no debris enters the pipe; they must also constantly verify that safety standards are complied with, to avoid accidents.

Steel pipes, fitted with flanges

In mini and micro hydroelectric plants, steel pipes are used in the lower sections of the penstock for safety reasons, when the pressure head exceeds the pressure capacity of commercially available PVC pipes. For example, in Peru the pressure head is 150 m, usually in one section. The rest of the uphill sections are PVC.

The steps below are a useful guideline:
- Clear the strip of land for the steel penstock section.
- Draw and stake out the centre line of the penstock, ensuring that it forms a right angle with the wall adjacent to the power house; check the ground levels with the levels of the penstock profile on the plan.
- In the powerhouse, establish the inlet level of the penstock with respect to the level of the finished floor and the turbine distributor.
- Locate and lay out the support bases and the distances between them.
- Locate the anchors of the steel pipe section, ensuring their alignment with the supports and the location of the expansion joint.
- Excavate the ground to form the platform for the supports, to the pre-established levels.
- Stake out the measurements of the support foundation for excavation purposes, fixing key and referential levels, in accordance with the penstock profile.
- Excavate the support foundation to the depth indicated in the penstock profile plan.
- Form and pour the concrete on the supports to the indicated height, having fixed the anchor bolts for the braces in the specified supports.
- Strip the forms off the supports and check the levels of contact with the penstock.
- Fit the pipe entering the powerhouse and the curve that changes direction in the anchor near the powerhouse, using provisional supports and the recommendations for fitting the pipes with flanges and accessories (washers, bolts, etc.).
- Continue working upwards, fitting the pipes on the built supports which should have set for at least 15 days and determine the final levels (contact surface between the pipe and the support).
- Fasten the piping to the supports and tighten the bolts of the clamps. B efore building the anchor near the powerhouse, check the alignment of the pipes.
- Build the anchor near the powerhouse, having previously completed the layout, stake out, excavation and location of the foundation level and verified the quality of the soil. Fix the curve so that it is not shifted by the pressure of the fresh concrete, formwork or casing (it must have set for at least 15 to 20 days).
- Complete the fitting of the pipe, working upwards towards the second anchor, including the installation of the expansion joint.
- Build the second anchor, bearing in mind that the curve (steel) that changes the direction of the penstock will be fitted into two ends of the steel pipes: the first downwards towards the pipe that will be coupled to the pipes being installed in the expansion joint and the second upwards towards the PVC penstock that will be coupled to an expansion coupling.

Before building the second anchor, the expansion joint and its accessories should be placed in the pipe going upwards, near that anchor.

The installation of the curves must be done with the best possible precision to ease the installation of the pipes in the established direction, preventing their misalignment.

- As in the installation of PVC pipes, the supervisor and the work force must ensure that no debris enters the pipes and that safety standards are complied with to avoid accidents and undesired setbacks.

Repairing PVC penstocks

Pressure pipes have been repaired on two occasions in the experience of the authors: the case of MHS Yumahual with a 8" RU Class 5 penstock (48 m head) and the case of Sondor with a 12" RU penstock (135 m head).

The penstock in Yumahual broke as a result of heavy rain during the El Niño phenomenon (1998), when part of the penstock was exposed and the impact of a heavy rock broke one of the penstock sections.

In Sondor, the cause was a leak and excess pressure in a socket joint. At that particular point in the penstock the pressure head was 112 m. The leak in the union was caused by a faulty pipe which had a deformation in one end: its cross-section was eliptical instead of circular. The supervisor thought that adding more PVC glue would be enough to fill the gap and create a strong joint as happens under normal conditions.

No leaks appeared during the commissioning tests, but after three months in operation it exploded at the leakage point, because it could not support the excess pressure in the penstock.

In both cases, the penstocks were repaired using the materials shown in Table 8.4 and procedure below.

Procedure: (see Figure 8.30):

Table 8.4 Materials used for repairing PVC pipes

MHS	Materials
Yumahual	Two 8" class 5 FU PVC repair unions
	Four 8" rubber rings
	One 8" class 5 PVC pipe
	200 g of lubricant
Sondor	Two 12" class 15 FU PVC repair unions
	1.50 m 12" class 15 PVC pipe
	Four 12" rubber rings
	500 g of lubricant

Note: 8" = 20.3 cm; 12" = 30.5 cm

- Make sure all the water is emptied out of the pipe, that the sluice gate next to the forebay tank is closed and that all the water is being diverted to the overflow channel.
- Examine the damaged part of the tube to discard it, bearing in mind that the unaffected pipe around it must have no cracks or any other flaws.
- Cut out the determined length of the pipe.
- Cut the new pipe to form an equivalent section to the damaged pipe that needs replacing.
- Clean the cut area of the installed pipe and the new pipe with sandpaper.
- Place the repair unions and the corresponding ring on either side of the installed pipe (Photograph 8.2).
- Attach the new pipe (section) between the cut pipe (installed) and mark the insertion length at both ends.
- Push each repair union up to the drawn mark, covering the clear space between the pipes.
- Carry out the stagnation test for at least 24 hours and put the pipe into operation.

Technical specifications

These refer to:
- the type of material selected (physical and mechanical characteristics);
- cutting the pipes;
- installation of the pipes;

Photograph 8.2 Fitting the repair unions into the damaged PVC penstock

PENSTOCK 143

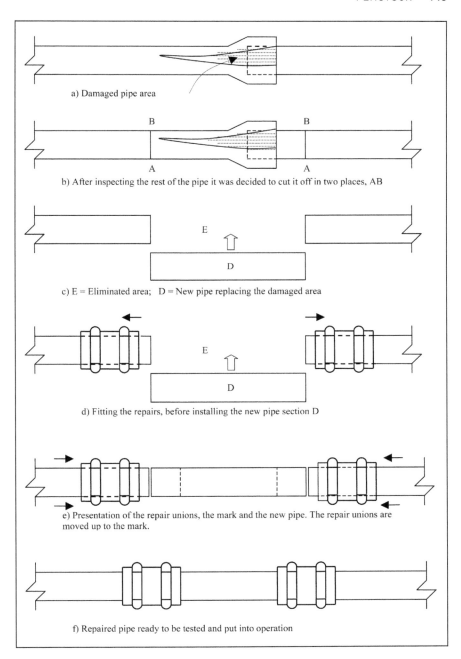

Figure 8.30 Pipe repair steps

- protocol tests;
- transport and storage.

The material can be either PVC or steel. In both cases, the technical specifications must indicate, for each section:
- resistance of the pipe to the pressure head;
- outer diameter;
- inner diameter;
- breadth;
- length.

Fitting the pipes

PVC

- Type of union between pipes: rigid (RU) with PVC glue, or flexible union (FU) with incorporated ring and lubricant.
- Minimum length of the fitting between the pipe end and the socket depending on the diameter.
- Fitting of the pipe and curves in accordance with the change of direction.
- Class or pressure rating of the pipe bends.

Steel

- Type of union between pipes: welded – type, shape and thickness of the weld. If standard flanges are used for the unions, indicate the type and thickness of the washer, the thickness of the bolts, screws, etc.
- Type of fitting of the pipe and the curve.
- Maximum and minimum separation between pipes in the area of the expansion joints.
- Location of the expansion joints.
- Physical and mechanical characteristics of the expansion joints.

Installation of the pipes

PVC

- Dimensions of the ditch (width and minimum depth).
- Thickness and material of the trench bed.
- Alignment of the pipes.
- Location and characteristics of the anchors.
- Curves manufactured for changing direction in accordance with the design profile.
- Covering the ditch.
- Embedding the pipes in the forebay tank.
- Minimum length and inlet level of the pipe in the powerhouse.
- Coupling of the vent pipe.
- Geology and bearing capacity of the soil.

Steel

- Characteristics of the supports (shapes, dimensions).
- Foundation levels.
- Bearing capacity of the soil.
- Distances between supports.
- Embedding the pipe in the forebay tank.
- Minimum length and inlet level of the pipe in the powerhouse.

Protocol tests

- Hydrostatic test: minimum and maximum application time, pressure range and length submitted to pressure, testing schedule. If testing equipment is not available then the hydrostatic test should be carried out for at least 24 hours.
- X-ray tests of the welded joints for rolled and electro-welded steel pipes.
- For supports and anchors, type and strength of the concrete, proportion of materials, minimum hardening time for the protocol test of the piping and fastening accessories.
- For the transport and storage of both PVC and steel pipes, follow the manufacturers' recommendations to avoid deformities that could make the installation difficult.

Table 8.5 Rieber union penstock systems, standard NPT ISO 4422

Nominal diameter		Rieber flexible union Length (m)		Series 6.6 (Class 15) Working pressure at 20°C: 15 bars			Series 10 (Class 10) Working pressure at 20°C: 10 bars			Series 13.3 (Class 7.5) Working pressure at 20°C: 7.5 bars			Series 20 (Class 5) Working pressure at 20°C: 5 bars		
mm	inches	Useful	Total	Inner diam. mm	Wall width mm	Approx. weight kg/U	Inner diam. mm	Wall width mm	Weight kg/U	Inner diam. mm	Wall width mm	Weight kg/U	Inner diam. mm	Wall width mm	Weight kg/U
63	2	5.90	6.00	54.20	4.40	7.32	57.00	3.00	5.13	58.40	2.30	3.99	59.80	1.60	2.83
75	2.5	5.89	6.00	64.40	5.30	10.48	67.80	3.60	7.32	69.40	2.80	5.78	71.20	1.90	4.00
90	3	5.89	6.00	77.40	6.30	14.96	81.40	4.30	10.50	83.40	3.30	8.18	85.60	2.20	5.56
110	4	5.88	6.00	94.60	7.70	22.35	99.40	5.30	15.81	102.00	4.00	12.13	104.60	2.70	8.34
140	5.5	5.87	6.00	120.40	9.80	36.06	126.60	6.70	25.21	129.80	5.10	19.42	133.00	3.50	13.48
160	6	5.85	6.00	137.60	11.20	47.28	144.60	7.70	33.42	148.40	5.80	25.58	152.00	4.00	17.96
200	8	8.84	6.00	172.00	14.00	73.88	180.80	9.60	52.09	185.40	7.30	40.24	190.20	4.90	27.53
250	10	5.81	6.00	215.00	17.50	115.44	226.20	11.90	80.75	231.80	9.10	62.71	237.60	6.20	43.52
315	12	5.77	6.00	271.00	22.00	182.89	285.00	15.00	128.25	292.20	11.40	99.01	299.60	7.70	68.15
355	14	5.75	6.00	305.40	24.80	231.07	321.20	16.90	161.23	329.20	12.90	124.52	337.60	8.70	85.01
400	16	5.74	6.00	344.00	28.00	293.92	361.80	19.10	205.28	371.00	14.50	157.72	380.40	9.80	107.90

Source: Factory catalogue, Amanco Del Perú SA

Table 8.6 Technical characteristics of PVC

Physical		Mechanical	
Specific weight	1.41 g/cm^3 at 25°C	Design stress	100 kg/cm^2
Water absorption	<40 g/m^2	Resilience to traction	560 kg/cm^2
Dimensional stability	150°C < 5%	Resilience to bending	750–780 kg/cm^2
Thermal expansion coefficient	0.06 mm/m/°C	Resilience to compression	610–650 kg/cm^2
Dielectric constant	A 103-106 Hz:3-3.8	Young's modulus	≈ 30,000 kg/cm^2
Inflammability	Self-extinguishable		
Friction coefficient	n = 0.009 Manning C = 150 Hazen-Williams		
Vicat point	≥ 80°C for aqueducts		

Source: Factory catalogue, Amanco Del Perú SA

Table 8.7 Rigid PVC pipes

Nominal diameter (ASTM)	Outer diameter (mm)	Class 15 (215) RDE-14.3 thickness (mm EC)	Class 10 (145) RDE-21 thickness (mm EC)	Class 7.5 (108) RDE-27-7 thickness (mm EC)	Class 5 (72) RDE-41 thickness (mm EC)	Length includes EC (m)
2"	60	4.2	2.9	2.2	1.8	5
2½"	73	5.1	3.5	2.6	1.8	5
3"	88.5	6.2	4.2	3.2	2.2	5
4"	114	8.0	5.4	4.1	2.8	5
6"	168	11.7	8.0	6.1	4.1	5
8"	219	15.3	10.4	7.9	5.3	5
10"	273	—-	13.0	9.9	6.7	5
12"	323	—-	15.4	11.7	7.9	5

Source: Factory catalogue, Interquimica SA (Peru)

Table 8.8 Steel piping systems

Nominal measures Inches	Schedule 40				Schedule 80				Schedule 160			
	Outer diam. inches	Wall thickness inches	Inner diam. inches	Weight lb/foot	Outer diam. inches	Wall thickness inches	Inner diam. inches	Weight lb/foot	Outer diam. inches	Wall thickness inches	Inner diam. inches	Weight lb/foot
2	2.375	0.154	2.067	3.66	2.375	0.218	1.939	5.03	2.375	0.343	1.689	7.45
2½	2.875	0.203	2.469	5.80	2.875	0.276	2.323	7.67	2.875	0.375	2.125	10.00
3	3.500	0.216	3.068	7.58	3.500	0.300	2.900	10.30	3.500	0.437	2.626	14.30
3½	4.000	0.226	3.548	9.11	4.000	0.318	3.364	12.50	4.000	—	—	—
4	4.500	0.237	4.026	10.80	4.500	0.337	3.826	15.00	4.500	0.531	3.438	22.60
5	5.563	0.258	5.047	14.70	5.563	0.375	4.813	20.80	5.563	0.625	4.313	33.00
6	6.625	0.280	6.065	19.00	6.625	0.432	5.761	28.60	6.625	0.718	5.189	45.30
8	8.625	0.322	7.981	28.60	8.625	0.500	7.625	43.40	8.625	0.906	6.813	74.70
10	10.750	0.365	10.02	40.50	10.750	0.593	9.564	64.40	10.750	1.125	8.500	116.00
12	12.750	0.406	11.938	53.60	12.750	0.687	11.376	88.60	12.750	1.312	10.126	161.00
14 OD	14.000	0.437	13.126	63.30	14.000	0.750	12.500	107.00	14.000	1.406	11.188	190.00
16 OD	16.000	0.500	15.000	82.80	16.000	0.843	14.314	137.00	16.000	1.562	12.876	241.00
18 OD	18.000	0.562	16.876	105.00	18.000	0.937	16.126	171.00	18.000	1.750	14.500	304.00

Source: Factory catalogue, Metales Andinos SA (Peru)

CHAPTER 9
Electromechanical equipment

This chapter contains a summary description of the components of the electromechanical equipment, the considerations and criteria borne in mind to select them and the general guidelines and technical specifications required to order the manufacture or purchase of such components. In other words, what is required by an engineer or field technician who wants to install electromechanical equipment in a small hydroelectric plant. The intention is not to design this equipment or focus on the theory of the operating principles of these machines; instead, the engineer or project technician is advised to obtain information from well-known or specialized manufacturers when selecting the equipment, because they could specify whether the equipment required is available on the market and could also provide technical details, such as the quality of the materials employed, etc.

The following are the main components of the electromechanical equipment:
- The turbine, which starts operating when the water intake valve on the penstock is opened, transforming the pressure energy and fluid speed into mechanical energy, making the shaft spin.
- The generator, which transforms the mechanical energy provided by the turbine into electricity.

In addition to these main components, there are complementary ones such as the load controller (which protects the power generator from damage caused by overspeeding if the load is suddenly reduced) and the transformer (which raises the voltage produced by the generator to ease the transmission of electricity).

Hydropower turbines

Hydropower turbines are classified into two large groups, impulse turbines and reaction turbines. Impulse turbines are those in which the water impacts the blades under atmospheric pressure. In this case, the water is directed towards the blades through an nozzle, which converts the kinetic energy of the water into mechanical energy. In reaction turbines, the water reaches the turbine blades at a higher pressure than the atmospheric pressure, but also at high speed; that is, it enters the turbine blades containing kinetic energy and pressure energy, which the turbine then transforms into mechanical energy.

However, a common and practical way of classifying hydro turbines is in three groups, depending on the relative head and flow: hydraulic turbines for high head and low flow applications (only impulse turbines are included in this group); medium head and medium flow turbines (both impulse and reaction turbines are included in this group); and low head and high flow turbines (this group only includes reaction turbines). (The terms high and low are used relatively; a low flow for a 1,000 MW power plant is very different from a low flow for a 1 MW power plant. Therefore, in order to judge whether a plant is within one of the ranges mentioned above, first of all the range of power must be identified.) A fourth group of turbines is currently being designed, also known as river turbines, which will work at zero physical head and operate using the dynamic head of the water corresponding to the speed of the flow.

All the models of turbines designed, except free current or river turbines, are within one of the three groups mentioned above. Some models could easily be placed in two of the three groups, particularly in the medium and low relative head ranges.

The classifications mentioned above are the result of the analytical and practical work of engineers who have specialized in the design and implementation of hydraulic power plants, in their attempts to find more efficient applications as well as a better cost–benefit ratio. For example, a Pelton turbine is only applicable in relatively high head and low flow conditions. If they were to operate outside this range, they would be inefficient and very heavy, because the higher the relative flow, the larger and heavier the turbine.

Whereas only the power generated or transmitted is taken into account for selecting elements such as the generator, controller and distribution grid, the turbine has a more variable performance depending on the site conditions. Consequently, for the selection of the electromechanical equipment, the turbine is the main element that requires attention.

It is important to select the turbine carefully as this will undoubtedly affect the costs, versatility and performance of the hydroelectric power plant. Technical-economic calculations are required for the selection of the other components (generator, controller and cables).

High-head low-flow turbines

Within this group, the two most well-known turbines are the Pelton and the Turgo type turbines.

Pelton turbine

This is the oldest and most widely used model of turbine in the world. It is operated by the impact of a water jet hitting the runner blades. According to the history of hydroelectric power plants, when a Pelton turbine was installed in the Fox River near Appleton, Wisconsin, on 30 September 1882, electricity had only been generated until then based on steam produced by carbon combustion.

The Pelton turbine is a very reliable, well-designed and robust machine. It is highly efficient and, unlike the majority of other models, it is also characterized for its high efficiency when working with partial flows. Micro, mini and pico Pelton turbines are currently manufactured, with one jet or with multiple jets depending on the head and flow conditions.

Turgo turbine

This is an impulse turbine, very similar to the Pelton turbine, except that it is designed to receive a larger jet, therefore a larger flow hits the blade. Also the position of the jet relative to the bucket is different. Given the characteristics of its design, it can work with relatively lower heads than Pelton turbines in similar flow conditions. This type of turbine competes with the slower Francis turbines and cross-flow turbines.

Medium-head and medium-flow turbines

As their name implies, these are applicable to such conditions. The most well-known turbine in this field is the Francis turbine.

Francis turbine

In the case of the Francis turbine, there are varied designs which allows a large performance range. In other words, different shaft speeds can be obtained for a specific head and flow, depending on the design of the blades. This is because it is a reaction turbine – unlike the Pelton turbine, in which the shaft speed depends only on the diameter and net head. Reaction turbines work with a full inlet casing which is coupled directly to the radial inlet of the turbine; in some cases, small area reducers are used.

They usually have a set of guide vanes which can vary the flow depending on the energy requirements.

Cross-flow turbine

The most well-known of these is the so-called Michell-Banki turbine, in honour of its inventor (Michell) and a scholar who improved the design (Banki). This is an impulse turbine that spins as a result of the impact of the water jet on its blades. Unlike the Pelton and Turgo turbines, this one has a larger rectangular type nozzle and the blades are built in such a way that the jet impacts the blade. The runner is shaped like a drum formed by a group of blades welded onto two parallel discs. As will be explained below, higher specific speeds can be obtained with this machine than with a Pelton or Turgo, operating with the same head, therefore larger flows can enter it.

A Michell-Banki turbine can usually replace a Francis turbine, the advantage being that its construction is much simpler. However, these machines are less efficient and will not last as long, although the duration can be improved if the manufacturing materials are chosen with care and the right manufacturing processes are followed. It is also worth mentioning that turbines of this type have often been used in micro hydroelectric schemes, particularly because they require simpler manufacturing facilities than are required for other types of turbine. For example, this type of turbine does not require casting, whereas the Pelton and Francis turbines do.

Pumps as turbines

In recent decades, there have been persistent suggestions to use rotodynamic pumps as turbines. Laboratory tests show that these machines can work efficiently when used as turbines (i.e. pumps in reverse). There have been some practical experiences in various parts of the world.

The most important advantage of selecting a pump is that these can be easily found in the market ready to be installed and used, whereas turbines have to be ordered and manufactured with technical specifications. The pumps most recommended are centrifugal pumps with higher specific speeds. The cost of a pump is usually lower than the cost of an equivalent turbine and they can be delivered off the shelf.

Low-head, high-flow turbines

These are normally used when river gradients are low and can be found in areas with less rugged geographical features. The turbines used in these cases are axial turbines. Like the Francis turbines, these are also reaction turbines in which different shaft speeds can be obtained for the same head, depending on the positioning of the blade angles. A wide range of models has been designed and installed, depending mainly on the installation conditions and resources.

Kaplan and propeller-type turbine

Kaplan and propeller-type turbines are similar to a ship's propeller, except that when used in a turbine the force of water makes them spin to produce energy, whereas in ships the opposite occurs.

Kaplan and propeller-type turbines are designed and manufactured under the same principles and concepts, the difference between them being that Kaplan turbines have adjustable blades that can vary position depending on the flow and head conditions and can therefore be applied efficiently in a wider range with respect to both parameters. In a turbine with adjustable blades, the position of the blades is constantly regulated automatically in response to the demand for energy.

One of the most frequent problems encountered in the implementation of hydroelectric schemes is that turbines of this type are rarely manufactured in developing countries, whereas other kinds, particularly small impulse turbines, are manufactured commercially in many of these countries.

Other types of turbine

In addition to the various designs that can be obtained for each of these different kinds of turbine, there are alternatives which may be considered unconventional, such as the use of a ship's propeller in reverse. For another example, water wheels with large transmissions are also used to generate electricity. There are many examples of pumps being used as turbines, as well as laboratory tests that prove that these can work efficiently as power generators (see Sánchez, 1988).

Selection of the appropriate type of turbine

As mentioned previously, every kind of turbine requires adequate head and flow conditions to achieve the best performance. Experienced designers of hydroelectric plants can easily identify the type of turbine that should be used for specific head and flow conditions, although they would still have to prove it with a few simple calculations. Designers with little or no experience may find it difficult to identify the most appropriate turbine at first sight, therefore they may need some parameters to guide them in the selection of the machine so as not to make mistakes or take too long using the trial and error method.

For this reason, some techniques have been developed over time for quick selection purposes, depending on the physical conditions of the installation site (head and flow). Figures and tables have been prepared showing the type of turbine and its field of application, both tools can be used.

Figure 9.1 is commonly used to select turbines, particularly when they are for mini, micro and pico hydroelectric plants. The corresponding flow and head for different types of turbine are shown on the graph, using logarithmic relations between the parameters.

The flow in cubic metres is found on the horizontal axis and the head in metres is found on the vertical axis. For any situation, the corresponding type of turbine can be selected based on these two parameters. Every kind of turbine

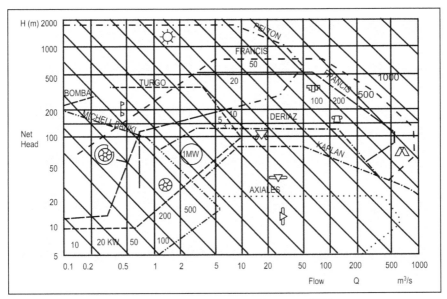

Figure 9.1 Graph for the selection of the type of turbine for a MHS
Source: Marchegiani, 2006

154 DESIGNING MINI AND MICRO HYDROPOWER SCHEMES

has a working area in which it performs adequately in terms of efficiency as well as versatility and cost engineering.

When the area of one type of turbine overlaps the area of another this means that either of the two turbines can be chosen. When that occurs the designer should look at other criteria such as costs and availability in making the final choice.

Turbine specific speed

What matters to a hydroelectric plant designer is how the turbine will perform in specific site conditions of head and flow and when there are variations in either of these parameters. The flow is more likely to vary during the year due to different rain patterns for the different seasons and it may affect the flow entering the turbine. On some occasions the head may also vary: for example in steel pipes the net head tends to be smaller due to the increase of friction losses in the pipe when it ages. It is also important for the designer to know how the turbine would perform if it was installed at a site with a different head from that for which it was designed.

To find out the technical characteristics of the performance of a turbomachine, or a turbine in this case, it is necessary to carry out a number of controlled tests so that the parameters can be measured accurately. It is also necessary to simulate the operation of these machines in a wide range of each of the above-mentioned

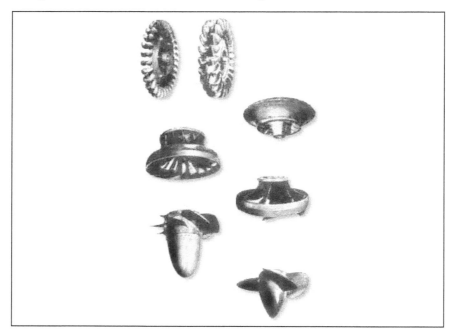

Figure 9.2 Different types of turbine (clockwise, from top left): Pelton, Pelton, Francis, Francis semiaxial, Kaplan, Kaplan, Francis

parameters. The only way to do this is through the turbine law of similarity. The law of similarity establishes that a large turbine used in a large power generating system performs in the same way as its laboratory counterpart; consequently, nowadays one can resort to graphs based on laboratory experiments that predict the efficiency and output of the machine to be used, based on similarity parameters.

The most frequently used parameters of the laws of similarity are:
- Specific speed based on power:

$$N_s = \frac{N\sqrt{P}}{H^{5/4}} = N\frac{\sqrt{(\rho g Q H \eta)/K}}{H^{5/4}}$$

- Specific speed based on flow:

$$N_q = \frac{N\sqrt{Q}}{H^{3/4}}$$

where:

N_s and N_q are the specific number of revolutions of power and flow, respectively

N is the rotation speed of the turbine in revolutions per minute (rpm)

P is the power of the turbine (HP or kW)

Q is the design flow of the turbine (m³/s)

H_n is the net head

Although both parameters can be used, the use of N_s is more common for hydraulic turbines.

Experimental work and engineering tests have produced a significant number of graphs and tables related to the similarity of the turbines, which make it easier to select the turbine for each case. Tables 9.1 and 9.2 contain a summary of the ranges of application of the different turbines, depending on their power specific speed (units in m, kW and rpm).

Observations and recommendations regarding the use of specific speeds for turbine designs

- Values in Table 9.1 relate to SI units (m, kW, rpm). However, this parameter is neither non-dimensional nor is it a speed, therefore some authors

Table 9.1 Hydraulic turbines and specific speeds (SI)

Type of turbine	Specific speed range (N_s)
Pelton	4–26
Turgo	20–56
Cross-flow	20–170
Pumps as turbines	30–170
Francis (spiral casing)	45–300
Francis (pit)	255–300
Kaplan and propeller	255–800

Source: UNIDO, 2004

Table 9.2 Main characteristics of hydraulic turbines

Turbine	Inventor and patent year	N_s	Q m³/s	H m	P kW	η_{max} %
Impulse						
Pelton	Lester Pelton (USA) 1880	1 Jet 25 2 Jet 25–50 4 Jet 25–50 6 Jet 43–60	0.05–50	30–1,800	2–300,000	91
Turgo	Eric Crewdson (Great Britain) 1920	50–225	0.025–10	15–300	5–8,000	85
Michell-Banki	A.G. Michell (Australia) 1903 D. Banki (Hungary) 1917–1919	34–138	0.025–5	1–50 (200)	1–750	82
Reaction						
Rotodynamic pump	Dionisio Papin (France) 1689	25–147	0.05–0.25	10–250	5–500	80
Francis	James Francis (Great Britain) 1848	L: 50–130 N: 130–215 F: 215–345	1–500	2–750	2–750,000	92
Deriaz	P. Deriaz (Switzerland) 1956	50–345	500	30–130	100,000	92
Kaplan and propeller	V. Kaplan (Austria) 1912	250–690	1,000	5–80	2–200,000	93
Axial						
Tubular Bulbous Peripheral generator	Kuhne, 1930 Hugenin, 1933 Harza, 1919	250–690	600	5–30	100,000	93

Note: N_s: Specific speed; L: Slow; N: Normal; F: Fast

consider that the name is unfortunate and that it should be called a 'shape number', as this would best explain its nature. Some references use shape or type numbers instead of specific speeds.
- For Pelton turbines, many publications subdivide the numbers according to the number of jets, assigning a different N_s range depending on whether there are one, two, three, etc.
- The majority of references quote minimum power, flow and head limits. Pico turbines do not usually fit into this range, because pico turbine technology is relatively new. It is recommended in this book that turbine selection is not based on the minimum ranges quoted, but on specific speed as its use is even valid for fractional kW turbines.
- The turbines used more frequently for small schemes in isolated communities are the Pelton and cross-flow (Michell-Banki) turbines. Axial turbines with fixed guide vanes (propeller) are also being used successfully at present.
- Pumps as turbines are a good alternative when the flow is steady all year round or varies only slightly.
- The use of simple belt and pulley transmission systems is common in small plants, particularly for less than 100–150 kW of power. For larger plants, direct coupling systems are designed or gearing is used as a transmission system.

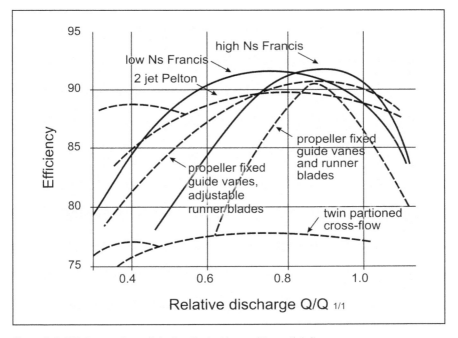

Figure 9.3 Efficiency of small hydraulic turbines with partial flows
Source: UNIDO, 2004

158 DESIGNING MINI AND MICRO HYDROPOWER SCHEMES

For a particular case, the final selection of a turbine is based on the following criteria.
- Characteristics of the site of the hydroelectric plant.
- Gross head (difference in levels between the water level in the forebay tank and the centre of the jet for action turbines or the level of discharge water for reaction turbines).
- Flow and head available for the plant.
- Rotation speed of the generator. In this case, the convenience of using mechanical transmissions (pulleys and belts or gearing) should be borne in mind.
- The probability of the turbine working with partial flows.

Figure 9.3 shows that, when operating with partial flows, efficiency of Pelton turbines is affected only a little, while efficiency in reaction turbines could be greatly affected. The graph also shows that, within the reaction turbines, different designs are affected differently when working at partial flows; for example the efficiency of low N_s Francis turbines is affected little by changes in flow, while the efficiency of high N_s Francis turbines is greatly affected when partial flows occur.

In practice stand-alone micro and mini hydropower plants are generally designed to work at full flow most of the time during the year (if not 100 per cent of the time). However designers interested in more information about the specifics of behaviour of turbines when working at partial load, should refer to in the literature in the bibliography.

To decide on the type of turbine to be used, the designer must consider the operating range of the turbine in terms of flow on the one hand and, on the other, evaluate the consequences of his or her choice regarding efficiency when the machine is working with partial flow. However, if the designer is sure that the flow will be steady all year, then partial flow behaviour of the turbine should not be a cause for concern.

Hydropower turbine sizing

Pelton turbine

Shaft speed

In Pelton turbines, the shaft speed is calculated as indicated in the following three steps:
1. Speed of the jet of water

    $V_{ch} = \varphi \sqrt{2gh}$

    V_{ch}, speed at the outlet of the nozzle (m/s)
    φ, speed coefficient (approximately 0.95–0.97), dimensionless
2. Tangential speed of the turbine
   The tangential speed of the Pelton turbine is calculated based on the speed of the incoming flow. (The tangential speed of the turbine is

produced for the PCD of the turbine -- the diameter at which the centre of the jet hits the runner.) In theory, the optimum tangential speed of the turbine should be half the speed of the incoming flow; however, experience shows that optimum performance speeds are slightly lower, therefore the following ratio is recommended:

$V_t = 0.45$ to $0.48\ V_{ch}$

The tangential speed can also be expressed as:

$$V_t = \frac{2\pi n r}{60} = \frac{\pi n r}{30}$$

From the above equation
3. Rotational speed of the turbine

$$\Leftrightarrow n = \frac{30 V_t}{\pi . r}$$

Where: V_{ch}, Flow speed at the outlet of the nozzle, m/s
g, gravity acceleration, m/s^2
h, net head, m
r, radius of the turbine, m

Using the data from Chetilla, Peru, and a commercial Pelton turbine with a 400 mm diameter, the following was obtained:

$V_{ch} = 46.10$ m/s, $V_t = 23.05$ m/s and rotational speed $n = 900$ rpm.

Using the units in Table 9.2 (m, hp, rpm) a value of $N_s = 32$ is obtained, which corresponds to a double jet Pelton turbine. The choice of a double jet turbine is appropriate because it allows for a better management of the flow, even if there are variations during the year. (If certain risks were taken when designing a hydroelectric plant with respect to the possibility of flow shortage during an extended dry season, it is important that the operator of the scheme is well trained so that when it occurs, he or she shuts down one nozzle to avoid a pressure drop in the penstock, which would seriously affect the generator.)

For a Pelton turbine, the most important dimensions and parameters required, in order to have a clear idea of its performance, are the speed and pitch circle diameter (PCD) of the runner, as well as the number and diameter of the jets.

PCD of the turbine

To estimate the PCD, the following is recommended, although equations directly related to the head may be used, as recommended in the bibliography.

$$D = \frac{60 V_t}{\pi . n}$$

As can be appreciated, the flow speed only depends on the net head, whereas the turbine diameter is based on the tangential speed and the angular or rotational speed and vice-versa (the rotational speed of the turbine will depend on its diameter). That is, in order to estimate the diameter of the turbine, it is

160 DESIGNING MINI AND MICRO HYDROPOWER SCHEMES

Table 9.3 Influence of the number of poles in the frequency

Number of pole pairs	Working frequency (Hz)	
	50 Hz	60 Hz
1	3,000	3,600
2	1,500	1,800
3	1,000	1,200
4	750	900
5	600	720

important to bear in mind the rotational speed and the type of transmission to be used, relating it to the rotational speed of the generator to be installed, since generators are manufactured to run at fixed speeds for either 50 Hz or 60 Hz frequency (see Table 9.3).

The generator speeds are fixed, therefore the value of n for the previous equation must match (or be very close to) one of those values in Table 9.3 when the generator is directly coupled to the turbine; otherwise this speed is divided by the transmission ratio to be used (when transmissions systems are used).

Diameter of the jet (d)

In hydroelectric plants with Pelton turbines, it is common to use more than one jet. The advantages of using multiple jets are:
- The diameter of the turbine is reduced when more jets are used, because a larger number of jets increases the flow intake area. With a smaller diameter, the machine will cost less and weigh less.
- A larger number of jets make it easier to regulate the flow between low stages and rainy seasons. This control consists of closing one or more of the nozzles during low stages when there is not enough water in the source (river), and keeping all the nozzles open during the rainy season.

The main disadvantage (although there are other minor ones) is that the losses in the distributor increase when the number of jets increases. There is no exact information regarding the ideal number of jets, as other factors intervening in the losses need to be analysed case by case. However, for small hydroelectric plants of less than 100 kW, it is advisable to use no more than three jets if possible.

The following equation was used to calculate the diameter of the jet (d):

$$d = 0.55 \left(\frac{Q}{\sqrt{H_n}} \right)^{1/2}$$

An important recommendation to bear in mind:

$10.0 \leq D/d \leq 11.0$

In practice, the diameter of the turbine, the diameter of the jets and the number of jets are estimated interactively, bearing in mind the aptness in terms of the rotational speed of the turbine, the transmission ratio and the desired number of jets. An examination of the previous equation shows that the value of d increases with an increase in the value of D, and therefore the number of jets required decreases. However, as mentioned in previous paragraphs, a higher D implies a greater cost and weight.

Another method is to decide on the number of jets and the flow distribution between them in advance. For example, a hypothetical case of 100 l/s can be distributed in three jets with a flow intake of 25 per cent, 35 per cent and 40 per cent, respectively, i.e. 25 litres, 35 litres and 40 litres per second. Based on that, the different diameters of the jets can be calculated using the previous equation. Once the diameters of the jets have been calculated, check whether they comply with the D/d ratio for the largest jet diameter as there will be no problem with the smaller ones. If the result is inadequate, the flow distribution can be varied to obtain the most appropriate value.

An alternative sizing method is to first select the number of jets and the flow distribution, calculate their diameters and then find the minimum PCD.

Cross-flow turbine (Michell-Banki)

The cross-flow turbine, commonly referred to by the name of its creators, Michell (Australian) and Banki (Hungarian), is an impulse turbine, therefore its operating principles are similar to the Pelton turbine. The main differences are the geometry of the turbine and the blades, as well as the way the water flow enters the turbine and the geometry of the nozzle.

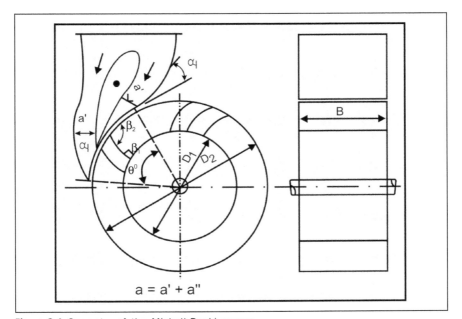

Figure 9.4 Geometry of the Michell-Banki runner

Geometry of the nozzle

In this turbine, the nozzle is a rectangular section comprising two straight sides that discharge the flow in the entire width of the runner (see Figure 9.4) and an enveloping upper side that guides the flow. This side has an optimum α_1 angle, constant at every point of the curve. The absolute flow speed will be tangent at every point of this curve.

However, the specific characteristics of the admission span, angles and other manufacturing details of cross-flow turbines are usually determined by the manufacturers, as they have pre-designed models that respond to the needs of the promoter or installer of hydroelectric plants of this kind.

In brief, the most important parameters required for sizing the equipment are the rotational speed and the diameter of the runner. Guidelines are therefore provided to this end.

The speed of the water when it leaves the nozzle is given by:

$$C_1 = k_c \sqrt{2gH_n}$$

where:
C_1, speed of the water entering the turbine
k_c, speed coefficient (≈ 0.95)
g, gravity acceleration (m/s²)
H_n, net head (m)

Rotational speed of the turbine

Most authors recommend the sizes most often used by the manufacturers (i.e. 100 mm, 200 mm, 300 mm…), as these are usually commercially available. A special size ordered by the buyer would probably be very expensive.

Once the diameter is selected, the rotational speed of the machine is calculated, using the following ratios:

$u_1 = 2.127 k_c \sqrt{H_n}$, tangential speed of the turbine (m/s)

This equation is appropriate for specific nozzle intake angles $\alpha_1 = 16°$ and $\beta_1 = 30°$; this selection corresponds to values calculated for maximum efficiency in different positions.

$$n = 40.62 k_c \sqrt{\frac{H_n}{D}}$$

Different authors allocate different values to k_c; however the majority suggest values of 0.95 or more.

Partial flow efficiency

Cross-flow turbines, like Pelton turbines, work well at partial flows; however when there are big changes in flow during the year designers tend to split the turbine runner in two sections across the width of the runner. As shown in Figure 9.5, the common way of splitting it is 1/3 and 2/3, each with an

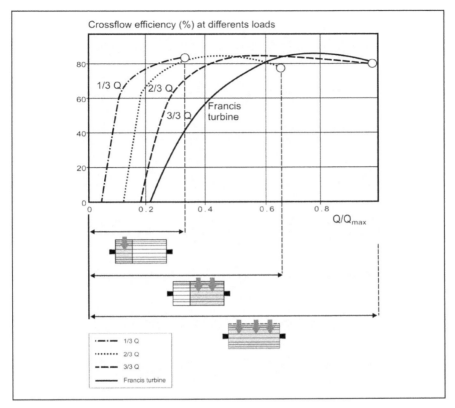

Figure 9.5 Efficiency of the Michell-Banki turbine at partial flows
Source: Marchegiani, 2006

independent nozzle. Therefore, the turbine can work with nozzle 1 only (with one third of the flow), with nozzle 2 only (with two thirds of the flow) and with both nozzles. In this way this machine can always maintain a high relative efficiency rate.

Width of the runner

The width of the turbine runner can be estimated by the following equation. This is an important parameter as it will give the designer of the hydroelectric plant an idea of the dimensions and weight of the equipment and other useful details for the installation of the system. However, the exact dimensions of the turbine will only be obtained when one is selected from among those existing in the market for turbines of this kind.

$$B = 98 \frac{Q}{D\sqrt{H_n}} \frac{1}{\theta°}$$

Where $\theta°$ is the turbine admission angle.

Table 9.4 Values of K_a, depending on the admission angle ($\alpha_1 = 16°$)

θ	60°	90°	120°
K_a	0.1443	0.2164	0.2886

Source: ITDG, Peru 1996

Thickness of the jet

The turbine admission angle bears a relation to the number of blades and the geometry of the nozzle. The following equation corresponds to that relation.

$a = K_a D$

Where a is the thickness of the jet, K_a is the dependence coefficient of the nozzle angle, θ is the turbine admission angle. For $\alpha_1 = 16°$

Pumps as turbines

There are a number of studies and practical experiences on the use of pumps in reverse (as turbines). Although centrifugal pumps are the most widely disseminated, in theory and according to laboratory tests, there are no restrictions on using axial and mixed flow pumps as turbines. Potential problems are the same as for other types of turbine with similar specific speeds. For example, there could be cavitation problems in fast axial turbines and runaway speeds are higher than those produced when centrifugal pumps are used as turbines.

The main advantages are:
- Standard pumps are commercially available, therefore manufacturing time and supervision efforts are avoided.
- The cost of a pump is lower than that of an equivalent conventional turbine, mainly because it is based on standard models, whereas turbines are made to order.
- Spares and technical assistance are available in the domestic market.

Some authors point out the following disadvantages:
- Their efficiency rate is relatively lower than a custom-made turbine, although in practice this is rarely the case, unless the competing turbine is highly finished and costly, which does not often occur in small applications.
- Their efficiency may drop suddenly at partial flows, as they have fixed vanes.

Francis turbine

Designing and selecting Francis turbines is hard work and project engineers find it difficult to make accurate estimates of their dimensions for manufacturing purposes. In practice, once the project engineer is convinced that a Francis

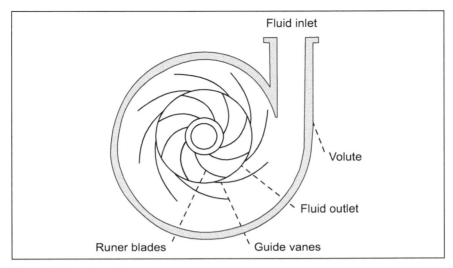

Figure 9.6 Francis turbine

turbine is the most appropriate alternative, he or she will take the flow and head details to the manufacturer and request a quotation.

Estimating the dimensions of a Francis turbine is more complicated than for impulse turbines because the fluid inlet conditions, particularly the speed, vary according to the design. This is done by varying the angles at which the water enters the vanes, which implies that the specific speed and finally the dimensions of the turbine must also vary. In practice, manufacturers tend to have models designed for micro and mini hydroelectric plants and they adjust them to the particular requirements of each case.

Axial turbines

The previous observations on Francis turbines also apply to axial, Kaplan and propeller turbines. It is advisable to provide the details to the manufacturer and obtain an offer in accordance with the designs available.

Electric generators

These machines receive mechanical energy from the turbine shaft and transform it into electrical energy. This transformation occurs by a magnetic field with lines of flux between the rotor poles and an arrangement of windings on the motor frame (stator) . Each winding is made up of a group of coils.

In an electric generator, the most important parameters are presented by the voltage (V) generated and the frequency (F). The latter parameter depends on the rotational speed of the electric generator.

The two large groups of generators are direct current generators and alternating current generators.

166 DESIGNING MINI AND MICRO HYDROPOWER SCHEMES

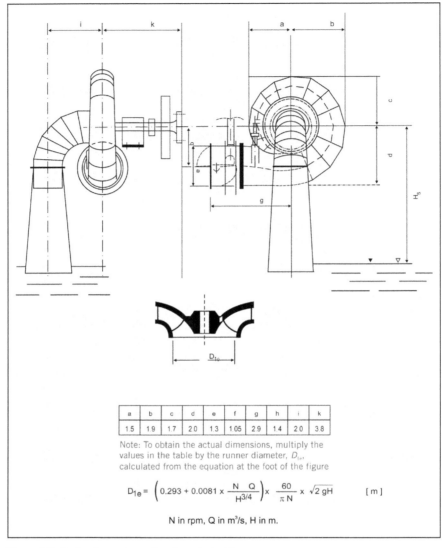

Figure 9.7 Basic dimensions of a Francis turbine

Direct current generators

These are no longer used in hydroelectric plants, although they were used when the power generating technology first developed. The most well-known is the conventional Dynamo generator commonly used during the initial application of hydroelectric plants.

Synchronous generators

These are the kind currently used for generating electricity in general. This generator has evolved significantly in terms of power magnitudes, structure, cooling methods and other factors that have made it the most appropriate power generating machine.

Asynchronous (induction) generators

This is the term used for motors working in reverse (motors as generators). The main characteristic of these machines is that they were designed to receive electric energy and convert it to mechanical energy. When they are used as generators, they operate in reverse.

This type of electric machine is currently used in very small power generating units. The main drawback is that because they were not designed to generate electricity, they need an initial excitation source (which is incorporated in the machine in the case of synchronous generators).

Therefore, in order to use a motor in reverse, a small unit must be designed and attached to the asynchronous generator, consisting of a capacitor or a group of capacitors that will provide the necessary electric current for the excitation and, consequently, start generating electricity.

Micro and pico hydroelectric plants and asynchronous generators

During the 1990s, motors were frequently used as generators in the smallest range of hydroelectric plants. In the case of pico plants (up to 10 kW of power), they were extraordinarily successful in reducing costs. Nowadays, a small asynchronous generator for a pico plant can be bought at a fraction of the cost of an equivalent synchronous generator.

Although recommendations in various texts suggest that motors could be used as generators up to about 30 kW of power, in practice very small machines of up to 8 kW or 10 kW of power provide the greatest advantages. Appropriate controllers have been designed for these small power ranges, referred to as IGC (induction generator controllers), which are also conveniently priced compared with the regulating costs of electronic controllers applied to synchronous generators.

It is also worth mentioning that the generators used in the car industry are also used occasionally in pico power plants, mainly because of their commercial availability and not because of their cost and efficiency, as they tend to be less reliable and less efficient than asynchronous generators.

A transmission system connects the turbine to the generator. A system of belts and pulleys can be used, which is very common in pico and micro hydro plants although rarely used in mini power plants.

The direct coupling system for the turbine and electric generator is generally used for small or larger power plants, although in small plants gearing transmission systems also tend to be used.

168 DESIGNING MINI AND MICRO HYDROPOWER SCHEMES

Control of hydroelectric plants

Appliances that are commonly used for both domestic and industrial purposes require a steady electricity supply, which means that the frequency and voltage must be as constant as possible. Depending on the applications, some appliances are more tolerant to variations than others. For example, for sensitive laboratory equipment, more control is needed than for an incandescent bulb. As a result, standards are established in each country specifying the variation tolerance in these parameters. This implies generating power with constant voltage and frequency values and a service that reaches final users with minimum losses.

There are two ways to proceed to generate power with constant frequency and voltage. The first is by regulating the water flow so that only the required quantity of water needed to produce the required quantity of energy enters the turbine. The second is by always generating power with a full load, so that part of the energy is used to meet the demand and the surplus goes to a ballast made of a group of resistances that limit the amount of current, heating either water or air.

The system most used in large plants is the flow controller, to which end hydraulic-mechanical governors are designed to detect minimum variations in the demand through frequency sensors, based on which they transmit orders to a valve (or guide vanes) adequately placed at the turbine inlet, which will open or shut, maintaining constant frequency.

The electronic load controller has a different principle; as mentioned previously, it diverts the excess electricity to a power dissipation system. Electronic load controllers are currently used in mini, micro and pico hydroelectric plants. Their main advantage is that they are much cheaper than hydraulic-mechanical controllers.

Examples of the selection of electromechanical equipment

Example 1

Selection of electromechanical equipment components for a micro hydroelectric power scheme in the Chetilla district of Cajamarca, Peru, with a current population of approximately 300 families. Physical conditions of the plant site: a gross head of 175 metres and a minimum flow of 80 litres per second, with a maximum demand estimated for 25 years of 75 kW. Chetilla is located in the central Andean highlands at an altitude of approximately 3,200 m asl.

Step 1. Selection of the turbine

The electromechanical equipment is selected once the design parameters are completely clear. That is, with the gross head and the corresponding piping, the net head is calculated ($H_n = H_b$ – losses due to friction and turbulence = 167 m) (see Annexe I, Example 10, calculation of the penstock for Chetilla). The minimum flow of the river or watercourse will depend on the type of structure used and the length of the channel; there will be small losses due to filtration and evaporation, therefore the minimum flow of the river (in this case 80 l/s) must be enough to feed the turbine, compensate the losses caused by filtration and evaporation and preferably leave a safety margin. If the design power of the system is not available, then it must be estimated (see Chapter 1). For this example, the desired electric power output is expected to be 75 kW.

Step 2. Calculation of the necessary flow to produce the required power

Using the equation $P = kQH$

In this case, as it is a micro hydroelectric plant, select a $k = 6.0$ value

Power $(P) = 75kW = 6.0 \times Q \times 167$

$Q = \dfrac{75}{6.0 \times 167} = 0.075 \text{ m}^3/\text{s}$

$Q = 0.075 \text{ m}^3/\text{s}$ which is equivalent to 75 l/s

With these values, revise the head-flow turbine selection diagram to select the most appropriate turbine.

In this case, if you draw parallel lines from the ordinates and the abscissas for the indicated flow and net head values, these lines will meet inside the area of the Pelton turbine; although this point is also within the area of cross-flow turbines (e.g. Michell-Banki turbine, see Figure 9.1). With these results, a Pelton turbine can be selected with confidence.

The same process is followed for the selection of any turbine, although Table 9.1 provides another alternative, to which end it is necessary to calculate the specific speed of the turbine, obtained with this equation:

$N_S = \dfrac{N\sqrt{P}}{H_n^{5/4}}$

In the case of impulse turbines, if the dimensions of the nozzle injector or nozzles are known, the specific speed will only depend on the net head and the water intake area. Likewise, the rotational speed will only depend on the net head the turbine will work with, as explained below.

Example 2

This example shows data on a micro hydroelectric power plant installed by the authors in 2000 as part of their work for ITDG, in Las Juntas, a town in the northern Andes of Peru.

Physical head 7.20 metres, minimum flow 1.2 cubic metres per second, electric demand 25 kW estimated for 25 years.

Using the equation $P = kQH$

In this case, the value of k must be moderate, as the power out is only 25 kW; therefore select a value of $k = 5.5$; likewise, the pipe will be very small, going directly from the forebay tank to the turbine, hence a head loss of 0.5 m (0.5/7.2) is estimated, which is equivalent to approximately 7 per cent of the losses in the pipe. This leaves a net head of 6.7 m.

$25 = 5.5 \times Q \times 6.7$

$Q = 25/(5.5 \times 6.7) = 0.678$ m³/s or equivalent to $Q = 678$ l/s.

With this result, we can consult the turbine selection graph (Figure 9.1) and enter 678 l/s (0.678 m³/s). From the figure we can see that the only option is the cross-flow (e.g. Michell-Banki) turbine,. However, see what happens when we verify it with the specific speed.

Taking the equation $N_S = \dfrac{N\sqrt{P}}{H_n^{5/4}}$ and considering that it would be coupled to an 1,800 rpm synchronous generator, then

$N_S = \dfrac{1800 \times \sqrt{25}}{6.7^{125}}$,

which gives a value of $N_s = 834.93$. As can be appreciated, this is a higher value than the scale shown in Table 9.1, which indicates $N_s = 800$ as a maximum; in other words, in theory there is no appropriate turbine for this case.

Tables and graphs are put together based on recommendations and on the experience of those who suggest them, which means that in theory, a Kaplan or propeller type turbine could be designed with the N_s value obtained; nevertheless, it is advisable to lower the turbine speed to acceptable values and that can be done with transmission systems.

For example, if a transmission system with a 1:2 ratio from the turbine shaft to the generator shaft is selected, you would have a turbine that only requires a speed of 900 rpm; if this value is replaced in the equation, then $N_s = 417.46$, a value perfectly consistent with Kaplan or propeller turbines.

As can be appreciated from the two previous examples, whereas the rotational speed of a Pelton turbine can be calculated if the PCD of the turbine and the net head are known, that is not possible with reaction turbines (Francis turbines, pumps as turbines or Kaplan and propeller turbines), because in addition to those parameters, the rotational speed of these turbines depends on the angle of the blade.

When selecting a slower propeller turbine in the case of Kaplan and propeller turbines, in addition to the reasons mentioned in the previous paragraph of not allowing N_s to be outside the recommended range, it is worth bearing in mind that the speeds of reaction plants usually have an impact on the useful life of the turbine. Turbines with a very fast specific speed will have high-speed fluid between their blades, which could result in a cavitation phenomenon. Alongside other reasons, it is advisable when working with reaction turbines to stay within the middle of the recommended range of N_s values to be certain of operating without cavitation.

In the actual case of the turbine in Las Juntas, the rotational speed of the turbine is 900 rpm and it is coupled to the generator by means of a transmission system of pulleys and belts, which brings the speed of the synchronous generator to approximately 1,800 rpm.

Technical specifications of a turbine for manufacturing purposes

Once the type of turbine has been selected, if the project designer has experience and is aware of the turbines available in the market, he can draw up a complete specification table to calculate and select the transmission system and the number of jets for Pelton turbines. Otherwise, it is recommended that, for the rest of the details, the project designer should work closely with the manufacturer and, in some cases, provide global figures – head (H_n), flow (Q) and type of turbine, and request quotations.

Technical specifications of the turbine for Chetilla

- Installation site: Chetilla district, department of Cajamarca, Peru
- Type of turbine: Pelton with two jets
- PCD: 400 mm
- Diameter of nozzles: nozzle A 'X' mm, nozzle B 'X' mm
- Distributor: built based on a Schedule 40 steel pipe
- Nozzles built in a 4" Schedule 40 pipe
- Manufacture: stainless steel turbine runner, manufactured in a single piece, with at least 24 buckets, thermally treated
- A special steel shaft with a 40 mm diameter
- Double row of sealed ball roller bearings
- Casing built of ¼ thick sheets
- Instruments: each nozzle should have a pressure gauge with a maximum column capacity of more than 300 m
- The distributor pipe must have a pressure gauge of the same pressure as the nozzles.
- Transmission: 1:2 ratio based on pulleys and belts, suitable for the power and torque conditions of the plant
- Minimum guarantee: 'X' years

Technical specifications of the turbine for Las Juntas

- Installation site: Las Juntas town, Pomahuaca district, Cajamarca, Peru
- Net head: 6.8 m
- Flow: 668 l/s
- Type of turbine: propeller (vertical shaft)
- Rotational speed: 900 rpm
- Manufacture: stainless steel runner manufactured in a single piece and thermally treated
- A special steel shaft with a 40 mm diameter
- Double row of sealed ball roller bearings

- Casing built of ¼ thick sheets
- Instruments: install a pressure gauge with a maximum pressure of more than 50 m head of water in the casing at the turbine inlet
- Transmission: 1:2 ratio based on pulleys and belts, suitable for the power and torque conditions of the plant
- Minimum guarantee: 'X' years

Sealing hydraulic turbines

The sealing of hydraulic turbines is unavoidable, and good sealing is extremely important to prevent water leakages through the joint shaft-casing. Poor sealing of hydraulic turbines involves the following potential risks:
1) Damage to the turbine bearings. Turbine bearings are generally located close to the casing, therefore leakages through the shaft-casing joint can reach them very easily and cause severe damage to the bearings.
2) Potential short circuits if water leakages reach any exposed electrical materials in the powerhouse.
3) Risks of accidents resulting from wet floors in the powerhouse.

See details regarding conventional and hydraulic seals in the final annex.

Electricity transmission and distribution

The electricity produced by the power generating equipment is transmitted to users via a transmission and distribution system. Hydroelectric systems usually require a high or medium voltage system to transmit electric energy from the generating point to the receiving system, which can be the national grid or the town to be supplied. There are a few exceptions when only a low voltage is required, which can occur in micro or pico hydroelectric plants, but not in mini hydroelectric power or larger power plants.

High and medium voltage cables can transmit electric energy for long distances with fewer losses; the higher the voltage of the system, the fewer the losses, therefore it is customary to use high voltage cables in large schemes. However, it is also worth clarifying that the higher the voltage, the more sophisticated and costly the grid, since the low/high and high/low voltage transformers are expensive. Similarly, the higher the voltage, the higher the pylons that are required and insulation and protection systems also need to be suitable for such working conditions.

Transmission systems for mini and micro power plants and even for pico power plants (not very common in the latter case) are medium voltage due to the short distance over which the energy is transmitted.

Distribution systems in small villages are usually low voltage (90 V to 240 V). However, this is only possible for very small populations, usually less than 200 to 300 families. For larger populations, a medium voltage distribution grid is required, and then low voltage distribution circuits, which deliver energy to final users. To transform medium voltage to low voltage power, an adequate number of transformers are used, depending on the size of the town.

CHAPTER 10
Powerhouse building

This is the structure that houses the electromechanical equipment and where the tailrace channel begins its course.

As mentioned in the previous chapter, in mini and micro hydro schemes the electromechanical equipment consists of the turbine, the generator and the electronic load controller. The control panel is an auxiliary component.

For the location, the following criteria are recommended:
- For the foundation, the ground on the site of the powerhouse and the area surrounding must be stable enough to withstand the weight equipment and the typical manoeuvres necessary for installation and maintenance of the system.
- If the river valley is curved, it should be placed in the convex part to avoid the sediment left in the concave part and prevent possible erosion by the river.
- The powerhouse must be located well above the level of the flow in the river during the rainy season, so that the tailrace channel can work properly throughout the year.
- The slopes of the penstock must not be prone to any subsidence, landslides or falling stones that could not only affect the structure of the powerhouse but the electromechanical equipment as well.
- There must be access to the forebay tank and the rest of the MHS components as far as the intake weir, as well as to the target population.
- The penstock must form a right angle with the wall of the powerhouse for good positioning of the electromechanical equipment.

Design

This involves designing the following:
- Architecture;
- structures;
- electrical installations;
- sanitary facilities.

Architecture

The architectural design usually consists of two rooms, one for the installation of the electromechanical equipment, called the powerhouse, and the other as a storeroom for tools, spare parts and the operator's quarters.

174 DESIGNING MINI AND MICRO HYDROPOWER SCHEMES

The dimensions of the powerhouse depend on the size and distribution of the components of the electromechanical equipment. It is also important to have enough room for the operator to move around. However, in order to keep costs low, it should not be too big.

The layout of the electromechanical equipment must be coordinated with the project designer as well as the designer of the low and medium voltage electrical system, so that the location of the cables and/or voltage transformer, earthing pit, etc. are well defined.

Example

In the 40 kW Yanacancha MHS, the powerhouse was designed as shown in Figure 10.1.

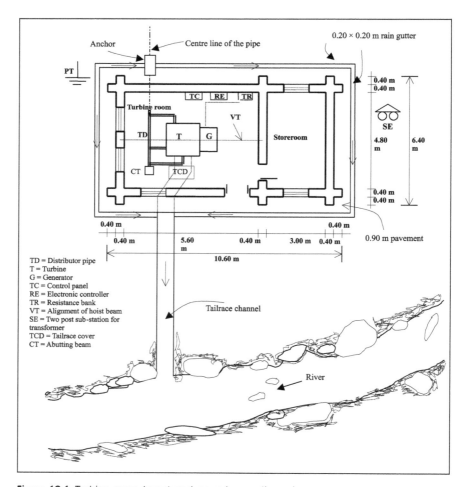

Figure 10.1 Turbine room layout and powerhouse dimensions

The main characteristics of the electromechanical equipment are as follows:

- A vertical shaft Pelton turbine with three jets. The transmission to the generator is via a system of pulleys and belts.
- The control panel and electronic load controller are assembled in a metal box which hangs on the wall.
- The resistance bank is placed in a tank on the floor, with a drainage gutter connected to the tailrace channel.
- There is a wooden hoist beam which is important for the assembly of the turbine and the generator and for maintenance activities.

In mini and micro hydroelectric plants, powerhouses are generally built with local materials, with adobe or mud walls, this being a reliable technology that people in isolated rural areas are familiar with. However if the soil in the

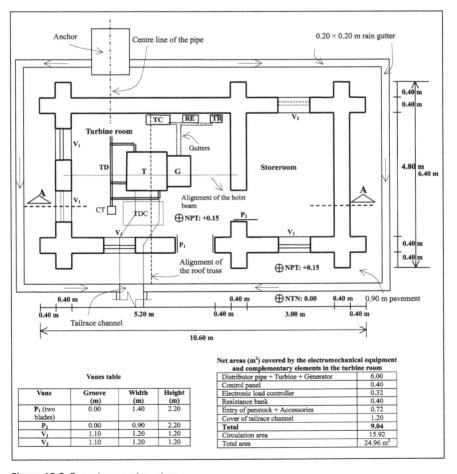

Figure 10.2 Powerhouse plan view

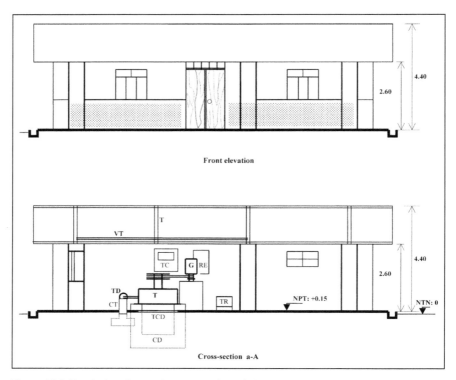

Figure 10.3 Front elevation and cross-section of the powerhouse

area is not suitable for building adobe walls, reinforced concrete columns and beams and brick walls are recommended. For example, in the Upper Jugle in Peru fringe areas, timber is the main component for walls.

Structure

The structural design consists of checking that the elements in the powerhouse are resistant to bending, compression and some seismic activity, especially the foundation, continuous footing, columns, beams and roof.

Figure 10.4 shows a a powerhouse with a coarse concrete foundation and continuous footing, adobe or mud brick walls and a wooden roof covered with roof tiles or corrugated zinc sheets.

POWERHOUSE BUILDING 177

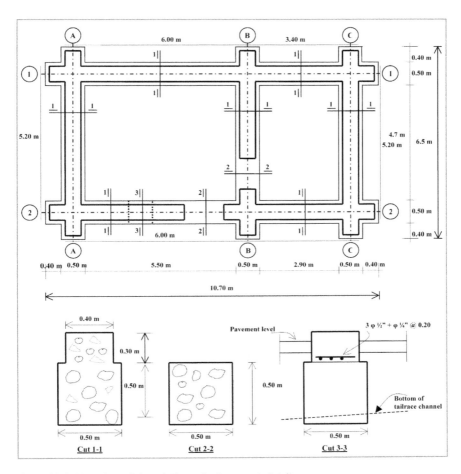

Figure 10.4 Plan view of foundations, footings and details

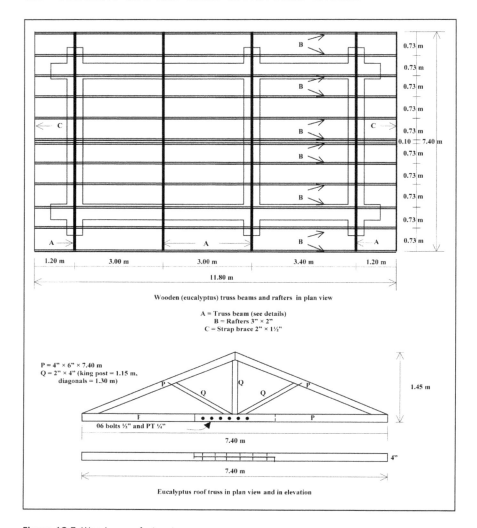

Figure 10.5 Wooden roof structure
Note: ¼" = 0.6 cm; ½" = 1.3 cm; 1½" = 3.8 cm; 2" = 5.1 cm; 3" = 7.6 cm; 4" = 10.2 cm; 6" = 15.2 cm

Example

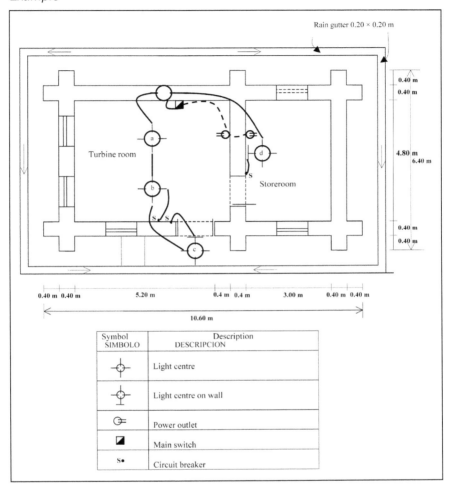

Figure 10.6 Electrical installation in the powerhouse (plan view)

Electrical installation

The design of the electrical wiring in the powerhouse is based on the lighting and power required for the operator's main activities.

Sanitary facilities

No special design for sanitary facilities is required as water is available to cover the operator's basic needs and to clean the powerhouse, taking advantage of the water that circulates through the resistance bank. To prevent bad habits and

180 DESIGNING MINI AND MICRO HYDROPOWER SCHEMES

contamination, a sanitary latrine or sanitary fixtures with a septic tank are required.

Technical specifications

In the powerhouse, the technical specifications refer to the four designs developed, depending on the construction technology employed. For example, the following are the technical specifications of this particular powerhouse.

Structures

Foundations

Coarse concrete foundation: volume mix C:Ag, 1:10 + 30% small stones. The size of the largest stone must be no more than 8" (20 cm).
Continuous footings of coarse concrete: volume mix C:Ag, 1:8 + 25% small stones. The largest stone must be no bigger than 5" (13 cm).

In both cases, the stones must be clean and of a good density.

The water to be used in the mix must be clean and free of any materials such as acids, alkali, salts, organic matter, clay, lime or other substances that could spoil the concrete.

Ditch water, sewage water or any water containing residues that could spoil the concrete must not be used to prepare and cure the concrete.

The concrete components must be mixed when dry, before adding water, until the materials are uniformly distributed. Enough water must be used to obtain a plastic mix, but avoid using too much water.

The foundation pit and the forming of the continuous footing must be cleared of debris and dampness before laying the concrete. The concrete must be transported and placed in such a way that its components will not segregate. It must only be transported in tins, wheelbarrows or buggies with rubber tyres and for permissible distances.

It should take at least seven days to cure the concrete in the foundation and continuous footings. Protect it from dry winds, heat or any other harmful factor. Curing water should be clean and the curing system carried out with wet sheets, or by dampening the concrete three or four times a day.

The foundations of the continuous footing should be strong enough to safely bear the load of its own weight and/or the pressure of the fresh concrete and/or the live loads acting in the foundation itself and while it is being applied. The foundation should not have any sagging or distortion and its surfaces must always be clean.

Adobe walls

Adobe walls must be stable to bear vertical and lateral loads, to which end the following characteristics are required:

Soil. Preferably clay soil with a low content of sand, free of organic matter and stones larger than 1/2" (1.3 cm) (small stones no more than 5 per cent). Bentonite must be discarded because it creates traction and contraction forces caused by humidity.

One way of identifying good soil is to prepare mud balls of approximately 1" (2.5 cm) diameter and let them dry in the shade for at least 24 hours, then try to break them by pressing a fist against each ball, applying force with the thumb and index finger. If it breaks easily, the material should be discarded. If it is difficult to break, it can be improved by adding vegetable fibre (*ichu* straw). If it cannot break, then it is good soil for adobe or mud bricks, bearing in mind that vegetable fibre must always be added.

Making adobe blocks. The block moulds must be manufactured first. The inner dimensions will depend on the size of the adobe blocks – usually the length and width of the adobe is the same as the width of the wall and its breadth (height) should be no less than 4" (10.2 cm).

Once the soil for making the adobe is selected, it must be crumbled, sifted and then soaked with water for one or two days, then beaten and mixed with chopped *ichu,* rice or barley straw (± 5 cm long and 1:8 straw:mud) to obtain the necessary plasticity for moulding. It is advisable to make some trial blocks.

To mould the block, a flat or level surface must be prepared to ensure that smooth adobes with no distortions are obtained. An adequate plasticity of the mud will prevent deformities when the blocks are removed from the moulds.

The adobe blocks should preferably be dried on both sides under shade, preventing a quick loss of humidity that could cause cracks. They must be stored in a safe place protected from dampness.

Adobe block bonding. The breadth of the wall must be taken into account when bonding adobe blocks. Before the first row and on the surface of the continuous footing, apply a 3:1 layer of mud and straw to bind the vertical and horizontal joints. It is advisable to dampen the surface of the adobe blocks slightly before putting them in place.

The walls and their corner joints or T joints must be squared, plumbed, levelled and set together with at least ¼" (6 mm) of the adobe.

Wooden crossbeams are placed at the indicated height of the adobe wall to hold the roof structure.

Door and window openings must be in the centre or at least 0.80 m from the corner.

Compacted mud walls

For walls made of compacted mud, the required soil is similar to that of adobe and the following technical specifications are required to build them.

Boxing work. No longer than 3 m, maximum height 40 cm, inner width at least 1/6 of the height of the wall (for one storey). The beam structure and other accessories must be aligned, plumbed, levelled, hard-wearing and strong enough not to be distorted by the strong impact of the compacted soil.

Preparing the soil. The soil must be sifted and free of any organic matter and stones larger than ½" (1.3 cm); preferably clay, cohesive and not prone to cracking once the damp has been eliminated. Add water only until the necessary degree of dampness is obtained, as well as vegetable fibre (chopped *ichu,* rice or barley straw) 5 to 8 cm long. Finally, blend the soil, straw and water until a uniform damp mixture that can easily be compacted is obtained.

Building the wall. This will be done with a wall box. Previously cover the opening that the boxing work left at the bottom with 3–4" (8–10 cm) flat stones. Place the prepared soil on the entire length of the wall up to a third of its height and compact it with a wooden mallet. Add more soil and compact it until it reaches the height of the wall box.

Continue with the rest of the boxes, following the same procedure for the first row, second row and so on.

Guarantee the stability of the walls, making sure that each box is aligned, plumbed and levelled and that the outside corners or joints are wedged to the inside walls with jambs. The length of the wall between jambs must not exceed 6 m.

Door and window openings must be in the middle of the wall and they must be as tall if not taller than their width. Finally, the roof structure will lean on the crossbeam at the end of the wall.

The timber for *the roof structure* must be dry, with no twisting, bending, warping or any other distortion and the dimensions of the trusses, rafters and straps must be strong enough to withstand their own weight, a live load and other stresses (wind, seismic forces, etc.) In areas with moderate rainfall, the slopes must be no less than 25 per cent and in areas with heavier rain between 30 and 35 per cent. The distribution of the roof trusses, roof coping, rafters, straps, etc. will depend on the type of roof (gable roof or hip roof).

Finishing touches

This refers to the quality of the floor, pavements, plastering, wall protection with external and internal skirting, the quality of the paint, the types of door and window, roofing materials, locks, etc.

The floor and pavements are usually concrete or sometimes stone slabs. The walls are plastered with mud with a sandy finish. Use water-based paint (distemper). Grooved wooden doors, wooden window frames with 2–2.5 mm transparent glazing. Outside skirting up to 0.90 m high and 1" (2.5 cm) thick, of cement, sand and ¼" (6 mm) gravel with a frosted finish. Inside skirting up to 0.20 cm high and 1" (2.5 cm) thick, with sand-cement mortar and a polished finish.

Zinc sheeting or roof tiles must be compatible with the roof structure.

Electrical installations

Indoor and outdoor electrical installations, which are only required for lighting and power outlets, are limited to the characteristics of the materials employed, such as: rectangular and octagonal plastic or galvanized metal boxes, a general thermo-magnetic 15 amp three-phase switch, a lighting circuit and power outlets with solid No. 14 wire. Appliances such as power-saving bulbs, radios, etc. must be 220/240 V.

Sanitary facilities

For the sanitary facilities, the technical specifications are limited to the recommendations for a sanitary latrine and, if there is a need to discharge sewage water to avoid contamination of the environment, a septic tank.

Procedure for building the powerhouse

Follow these steps:
- Clear the site of any brush or weeds.
- Draw the layout of the entire area, including pavements, a channel around the perimeter to drain the rainwater and a wider discharge channel.
- Shear and level the land to form the platform.
- Draw the centre line of the penstock on the prepared platform and put stakes in place.
- Check the gross height from the level of the water surface in the forebay tank to the level of the finished floor of the powerhouse, bearing in mind the type of turbine to be installed.
- Check the maximum river level in the rainy season and the bottom of the future tailrace channel.
- Lay out the foundations, making sure that the centre line of the penstock forms a right angle with the cross-wall. In the drawing, use survey poles to mark the centre line of the foundations and the width of the foundation and continuous footings.
- Excavate the foundation to the required design depth and discard the surplus material.
- Shape and square the dug out pit and verify the foundation level.
- Prepare the right mix of materials with a dry mix of cement and aggregate until uniform, then add the specified quantity of water and blend the mix with a spade until a pasty concrete mix with a good level of elasticity is obtained.
- In the foundation pit, mark the width of the tailrace channel and form the area to be crossed by it, or cover it with sandbags to prevent any additional work in digging the foundation.

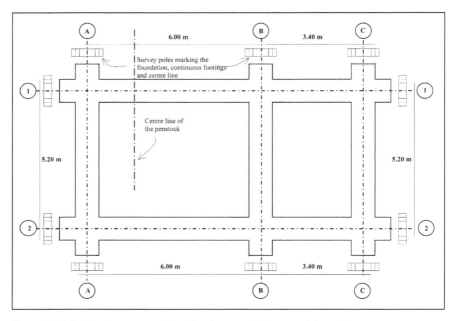

Figure 10.7 Putting in survey poles, drawing centre lines and foundations

- Dampen the foundation pit and then pour in the prepared concrete, putting large stones in place at the same time so that they are embedded in the concrete. While pouring the concrete, stir the concrete mass with a 5/8" (1.6 cm) iron rod or a 2" × 2" (5 cm × 5 cm) wooden rod, to compact it. Pour in enough concrete to reach the required level.
- 24 hours after pouring the concrete foundation, draw the base of the continuous footings on it, using the survey poles.
- Form the continuous footings, leaving the width of the door openings free.
- On the formwork, mark the height of the concrete for the continuous footings. Put a formwork support and the steel reinforcement in place to form a lintel over the upper part of the tailrace channel, where it crosses the foundation and continuous footings.
- Prepare the batch of cement and aggregate for the continuous footing; pour in the mixture and put medium stones in place, following the same procedure for the foundations.
- Cure the continuous footings for at least seven days.
- Fill in the wall with the selected type of masonry (adobe or compacted mud) in accordance with the technical specifications, making sure it is aligned, plumbed and levelled at all times. Temper or secure the corners and angles between the inside and outside walls, leaving the door and window openings free, with their respective ledges at the specified levels.

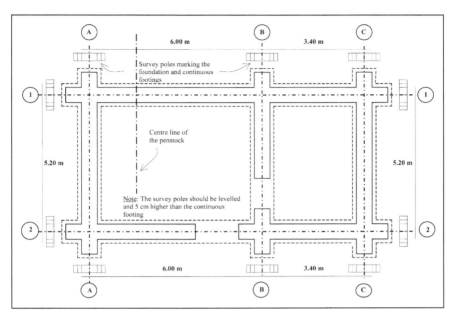

Figure 10.8 Alignment of continuous footings, using survey pole markings

It is also worth pointing out here that the point and level that the penstock will go through should be marked on the wall.
- Put the wall plate in place at the specified level, binding it to the ledges with binding wire. This beam must also be levelled to support the roof trusses or roof structure.
- Make the roof trusses with the help of a template previously drawn on a flat and levelled surface.
- On the wall plate draw the centre line of the roof trusses, verifying the distance between them and the corresponding level, bearing in mind whether the roof is a gable roof or a hip roof.
- Raise and attach the roof trusses on the wall plate, on the centre lines, ensuring that they are plumbed and separated by the distance indicated in the roof plan.
- Finish fitting all the other components of the roof structure, such as the ridge beam, rafter, straps, etc., always checking the distance and alignment for the selected roof.
- Put the selected roofing in place (corrugated sheeting or roof tiles), guiding the squared alignment with a rope and key points on each side of the roof. Then fit in the corresponding accessories that form part of the roof, trying to ensure that the longitudinal and crosswise lengths overlap. When fitting the roof tiles or corrugated sheeting, start in the lower vertex of the square, working up from the bottom and putting the covering on the ridge beam at the end.

- Put the hoisting beam in place in the position indicated in the plan in order to raise and assemble the turbine and generator and carry out other related activities.

This concludes the structural part. Next are the finishing touches, as follows:
- Plaster the inside and outside walls with mud and straw, to which end levelling and plumbing points must be placed on each of the walls, making the plaster no more than 1" (2.5 cm) thick. Work downwards from the top to the bottom.
- Sealing of door and window openings with mud and straw no more than 1" (2.5 cm) thick, plumbed and levelled.
- Construction of indoor skirting with cement and sand mortar in the proportion of 1:3, with a polished finish to make the lower part of the walls waterproof, particularly for powerhouse clearing and maintenance purposes. The skirting is usually 0.20 m high and 2.5 cm thick.
- Construction of outdoor skirting, usually with cement mortar, sand and ¼" (6 mm) gravel in the proportion of 1:2:3. The gravel gives it a rough finish. It is 0.90 m high from the pavement or at the level of the windowsill. Its purpose is to protect the lower part of the wall during the rainy season.
- Construction of floors and pavements. They are both at the same level so that during the cleaning and maintenance of the turbine room and the storeroom, the waste water will be conducted to the tailrace channel through the rain drainage gutter. The sub-base is built first, compacting the natural ground with or without surfacing and then building an 8 cm thick false floor with cement and aggregate in a 1:6 proportion. Then add a 2 cm layer of cement and sand in a 1:4 proportion and, for the final finish (polish), sprinkle with cement and smooth the surface with a plastering tool. The finished surface is the finished floor level.

The floor must not cover the site of the turbine, generator, tailrace channel and the gutter for the cables that will connect the generator to the control panel. This area must be left free with formwork, to ease the foundation of the turbine and generator, the connection of power cables inside gutters embedded in the floor and the construction of the inside of the tailrace channel. Also, make sure the rest of the floor is all one level to make it easier to assemble the turbine.

The pavement is built in a similar way to the indoor floor, except that it will have a 0.5 per cent slope towards the rain gutter and a float-finish with a perimetric and crosswise groove every 0.80 m.
- The doors are two different types. A double door is used at the front and a single door is used internally. The width of each door provides easy access for the electromechanical equipment to the turbine room and the storeroom before and during their installation. Both doors are made from local wood. They are usually made to order in carpentry workshops and the manufacturers themselves install them with their respective locks.

- The same goes for the windows, except that they have a security screen comprising ½" (13 mm) steel construction bars embedded vertically in the window mountings. The glass for the windows is made to measure and installed by a glazier.
- Before painting the inside and outside walls, sand down the plaster on the walls to eliminate the small cracks formed when the mud contracts. This consists of rubbing the wall with a damp rag with sand over the surface, then removing the dust and applying the paint. A seal coat of glue and paint must be applied first and then two coats of paint.
- For the construction of the rain drainage channel around the perimeter, use cement and aggregate in a proportion of 1:6. The gutter is 0.20 m deep and the walls are 0.10 m thick for the outer and inner wall. This gutter is built at the same time as the pavement. The bottom of the gutter is a concrete slab of the same proportion and 10 cm thick material with a two in one thousand slope towards the tailrace channel. Formwork is required for the construction to secure the required slope. The finish is semi-polished.
- The electrical installations are embedded. It is therefore necessary to patch the gutters made for the pipes and the recesses for the control panel, the cable box and the octagonal and rectangular boxes for the light centres, power outlets and switches, using gypsum plaster and a coat of paint.

Photograph 10.1 shows a typical powerhouse built by Practical Action in Peru.

Photograph 10.1 Powerhouse constructed with local materials
Note: Powerhouse built with concrete foundations and continuous footings, mud walls and a timber roof with roof tiles. Tailrace channel in excavation.

CHAPTER 11
Foundation for the electromechanical equipment

This chapter describes the foundation for vertical or horizontal shaft turbines and the generator base with a belt and pulley transmission system. The same criteria and procedure are used for schemes with direct coupling (turbine-generator).

Foundation for the horizontal shaft Pelton turbine (see Annexe Example 13)

The foundation for this type of turbine is a reinforced concrete structure, almost totally embedded in the floor of the powerhouse and joined to the foundation base of the generator. Its location is entirely related to the rest of the components and accessories.

The shape and dimensions depend on the measurements of the turbine and the generator, as well as other characteristics of the scheme, such as the easy evacuation of turbine water through the tailrace channel.

The characteristics of the foundation are as follows.
- The centre lines of the turbine and generator run parallel to each other and they both form right angles with the centre line of the penstock (Figure 11.1).
- The distance between the centre lines of the turbine and generator is provided by the project designer and/or the turbine manufacturer (Figure 11.1).

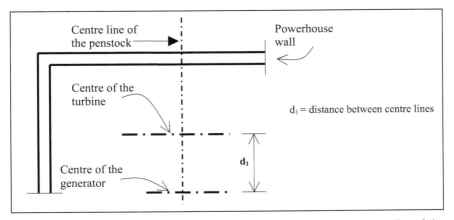

Figure 11.1 Centre lines of the turbine and generator with respect to the centre line of the penstock

- The direction of the water leaving the turbine (through one of the walls on the foundation base towards the tailrace channel) should not interfere with other connections of the electromechanical equipment (gutters, cables, etc.).
- The distance between the wall and the centre of the turbine depends on the accessories used to join the penstock to the turbine inlet pipe (see Figure 11.2).

In this case, the foundation fulfils two basic functions.

- It accurately and firmly holds down the metal base of the turbine on which the rest of the components will be assembled (casing, shaft, runner and accessories).
- It takes in the water from the turbine and carries it to the tailrace channel.

Design

Foundation

For the design of the foundation the supplier or whoever is responsible for the electromechanical equipment must provide the dimensions and weights of each component, their hydraulic conditions and relevant recommendations.

The foundation is a reinforced concrete tank with walls and floor of a specified thickness, with no top cover.

The inner dimensions of the foundation, L_i, A_i, are determined by the inner measurements of the metal base, which are the same in this case (see Figure 11.3).

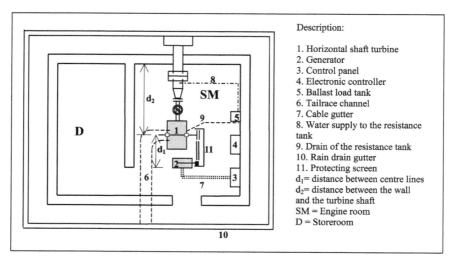

Figure 11.2 Locaton of the horizontal shaft Pelton turbine with respect to the rest of the components

The outer dimensions, L'_e, A'_e, are determined by adding the breadth of the walls and floor of the base, so that the anchors of the metal base have a coating around them of at least 7–10 cm.

The height, H'_e, includes the length of the anchor bolt (at least 20 cm) and should include the brace, free border and thickness of the channel floor, so that the runner can spin freely, away from the surface of the water in the sump that leads to the tailrace channel.

The base for the generator is determined in a similar way, taking the metal base as a guideline, except that instead of another tank it will be a concrete block with the same outer dimensions as the outer measurements of the metal base, plus a coating of at least 7 cm over the width of the anchors, bearing in mind the distance between the centre lines. This distance is provided by the supplier or designer of the equipment, otherwise the components should be pre-assembled as explained in the description of the construction process.

Calculation of the static and dynamic forces

The static and dynamic forces of the generator are created by:
- The weight of the equipment, the weight of the foundation and the weight of the live load.

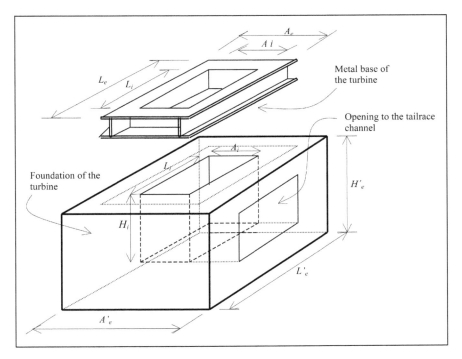

Figure 11.3 Sizing of the foundation, based on the measurements of the metal base of the turbine

- The forces produced by the rotation of the different generator components.
- The reaction of the turbine runner on the stator.
- The reaction in the generator caused by the electromagnetic faults (short circuits).

Admissible bearing capacity

The admissible bearing capacity of the foundation soil is calculated as follows.
Calculation of the actual weight of the reinforced concrete bases. Equivalent to the result of the volume multiplied by the specific weight of the reinforced concrete (2,400 kg-f/m^3).
Weight of the equipment. Weight of the turbine + weight of the generator and accessories.
Live load. Weight of the people and materials temporarily involved in the construction of the bases or the assembly of the equipment. In this case, 250 kg-f/m^2 was considered adequate.
Forces produced by the rotating motion. The forces produced by the rotation of the turbine and generator are described below.
Centrifugal forces are created by the eccentricity of the runner mass with respect to the shaft. This force is directly proportional to the output of the square of the angular speed × the runner mass × the indicated eccentricity. This centrifugal force is broken down into a horizontal force and a vertical force. In turn, these components create alternate horizontal and vertical vibrations, the effect of which increases the weight of the equipment and can reduce the resilience of the material if the quality is inappropriate.

Designers and manufacturers are responsible for providing equipment that fully responds to these and other forces that may be produced by the runner and generator. In practice, these forces are reduced to their minimum by dynamically balancing the runner and the generator. As per design criteria, it is recommend that, in order to absorb the forces caused by the eccentricity of the runner, nine times the weight of the elements affected by the rotation should be added to the weight of the equipment when the foundation structure is steel and 12 times if the structure is reinforced concrete.

For the forces mentioned above for horizontal shaft turbines, it is recommended that the foundations of both the turbine and the generator should be a single reinforced concrete structure to form a single unit with the factory structure (turbine and generator). This reduces the work coefficient of the static load and the danger of overturning or tilting.

The bearing capacity of the soil for the total static load should be less than the admissible bearing capacity of the foundation soil.

The foundation soil is usually near the river bank, geologically formed for many years and therefore stable, with a bearing capacity greater than 1 kg/cm^2; others even exceed 2 kg/cm^2. The soil is usually clay, with compacted sand, quarry stone gravel and, in some cases, clayey soil with stones larger than 12" (30 cm). This admissible bearing capacity must be greater than the load transmitted by the previously mentioned forces.

Construction process

Follow these guidelines:
- Extend the centre line of the penstock and put marks on the floor and on the wall (Figure 11.4)
- Lay out the connection between the turbine inlet pipe and the penstock (Figure 11.5).
- Lay out the connection between the inlet pipe with the nozzles and the casing of the turbine and generator (Figure 11.5).

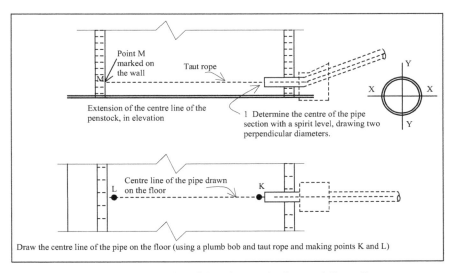

Figure 11.4 Layout of the centre line of the pipe on the floor and the wall

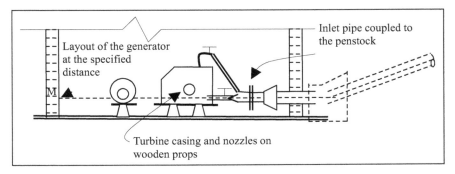

Figure 11.5 Profile of the layout of a horizontal shaft Pelton turbine

- On the ground, mark the centre line of the turbine QQ' and the generator PP' and the site of the turbine and generator. (Figure 11.6).
- Remove the layout ropes.
- Lay out the dimensions for the excavation of the foundation base (Figure 11.7).
- Dig the pit for the foundation of the turbine and generator to the specified depth, including the breadth of the paving. In some cases, larger than normal rocks are found during the excavation (Figure 11.8), which are so heavy that a 1 tonne hoist anchored to the hoist beam is needed to remove them, as well as other elements such as 5/8" to 1" (16–25 mm) ropes.

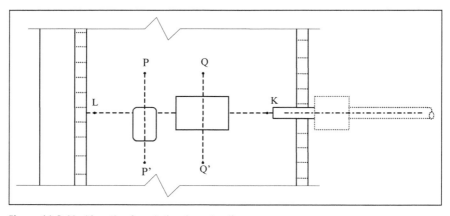

Figure 11.6 Marking the foundations' centre lines
Note: PP' = centre of the generator, QQ' = centre of the turbine

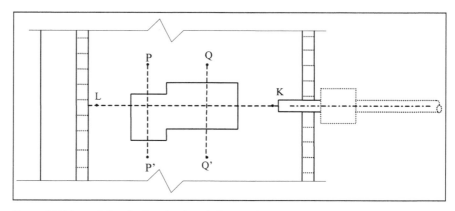

Figure 11.7 Layout for digging the foundation pit for the turbine and generator

FOUNDATION FOR THE ELECTROMECHANICAL EQUIPMENT

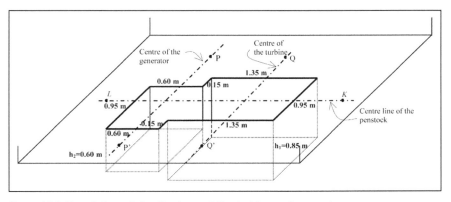

Figure 11.8 Foundation pit for the base of the turbine and generator

- Shape the foundation pit with appropriate tools and verify the foundation depth. When it is necessary to dig deeper than normal to extract large rocks, the profiling should be done with stone masonry or concrete.
- Compact the soil in the ditch.
- On the compacted soil, pour a 10 cm thick paving mix C:Ag 1:10 to make the foundation surface level and uniform. Put the level points in place prior to pouring this layer.
- Draw the centre line of the penstock and the centre line of the turbine on the paved bed, as guidelines for the steel reinforcement cage.
- Assemble and put the reinforcement cage for the turbine in place, using the centre lines drawn in the previous step as a reference. Use steel struts (½″ (1.3 cm) × 0.50 m rebar) to join it to the steel reinforcement cage for the base of the generator (Figure 11.9).
- Fix the steel reinforcement cage onto concrete blocks, trying to leave space free for a minimum coating of concrete.

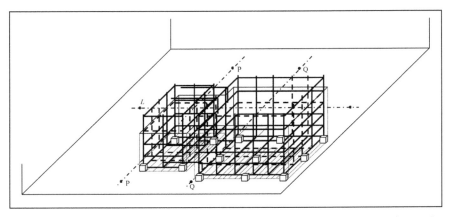

Figure 11.9 Fitting the steel reinforcement cage onto the concrete blocks, prior to the formwork

- Install the drain pipe that goes from the resistance tank to the pit, making sure when pouring the concrete that it neither breaks nor enters the pipe.
- Prepare the concrete mix and pour it on the floor to the specified height.
- Draw the centre lines of the penstock and the turbine again on the floor, as guidelines for forming the foundation walls.
- Form the foundation walls, making sure they are plumbed and squared to the required level, without forgetting the opening for the tailrace channel, in accordance with the specified width and height.
- Attach the metal base to the formwork, making sure it is levelled and aligned with the centre line of both the penstock and the turbine; in addition, check that the lower web of the channels (anchors) are free from the steel reinforcement cage.
- On the formwork of the walls, mark the level to pour the concrete.
- Prepare the concrete mix and pour it, applying all the recommendations regarding transport, compacting, etc. to the specified level, making sure the base did not shift during the process and that the anchors are embedded or completely covered with concrete.
- Clean the anchored base and the holes that will subsequently be used to bolt down the casing.
- Strip the form off the walls after a minimum of 24 hours.
- Render the floor and the walls in accordance with the recommendations and the final finish (do not place any weight on the walls until the concrete has been cured for the required period).
- Cure the concrete for at least 15 days before rendering.
- Follow the same steps to build the base for the generator, bearing in mind that it must be accurately aligned and levelled. The steel profile base of the generator, in addition to having holes to bolt down the generator, also has other elements for the adjusting screws. These must be kept clean and protected so that the screw threads are not damaged.

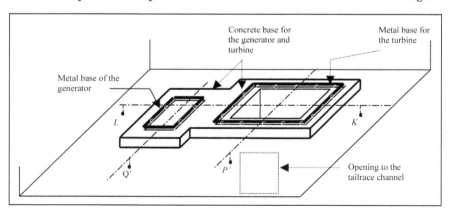

Figure 11.10 Metal base for the turbine and generator, anchored to the reinforced concrete foundation base

Technical specifications

This refers to the physical-mechanical characteristics of the materials used for the foundation and the admissible bearing capacity of the foundation soil.

In this case, the following materials were recommended:
- Concrete $f'_c = 210$ kg/cm^2 (volume dose $\to$ cement:sand:stones $\to$ 1:2:2)
- Steel: $f_y = 4{,}200$ kg-f/cm^2
- Coating of concrete over vertical and horizontal steel reinforcement bars. Minimum 2.5 cm
- Inner walls rendered with 3:1 sand–cement mortar, polished cement finish.
- Anchors: construction steel 5/8″ (16 mm) × 0.30 m electrically welded to the profile base.
- Metal bases, C profile (channel section) of structural steel 3/8″ (9.5 mm) thick.
- $\sigma_T = 2.00$ kg-f/cm^2 (verify on site).
- Outlet opening to the tailrace channel: (in this case) 0.40 wide × 0.60 tall

Other technical specifications that the turbine manufacturer considers advisable for the construction of foundation bases may also be included.

Photograph 11.1 shows the foundation base and the metal base anchored to it, for the turbine-generator unit corresponding to a horizontal shaft Pelton turbine.

Foundation for the vertical shaft Pelton turbine (see Annexe Example 14)

The foundation for this type of turbine is also a reinforced concrete structure embedded in the floor of the powerhouse and connected to the foundation base of the generator in the same way as the previous case.

The electromechanical equipment is laid out inside the powerhouse as follows.

The foundation of the vertical shaft Pelton turbine is joined to the foundation of the generator, taking up a larger area than the previous case.

Practically, the same technical criteria were applied to the layout of the turbine, generator and the rest of the components, to take the utmost advantage of the space and ensure an efficient circulation area.

- Based on the dimensions of the equipment, simulate a pre-assembly to scale, to determine the size of the occupied area and the best way to distribute it.
- Alignment with the centre line of the penstock.
- The centre line of the penstock is only aligned with the centre line of the turbine inlet pipe.
- The course of the centre lines of the turbine and the generator consist of two points X_1 and X_2 respectively, drawn on a straight section at right angles with the centre line of the penstock (see Figure 11.11).
- Two distances are evident: the first d_1 is between the centre line of the penstock and the position of the turbine shaft. The second d_2 is the horizontal distance between the shafts of the turbine and the generator.

Photograph 11.1 Foundation for the turbine-generator unit

FOUNDATION FOR THE ELECTROMECHANICAL EQUIPMENT   199

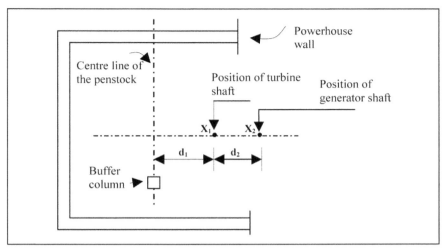

Figure 11.11 Plan view showing the location of the turbine and generator shafts with respect to the centre line of the penstock

- In this case, a reinforced concrete buffer column must be embedded in the ground at the end of the turbine inlet pipe.
- Direction of the tailrace channel outlet.
- The same criteria used in the previous case were considered, although in this case the turbine manufacturer recommended that an access should be left open in the tailrace channel (inside the powerhouse) underneath the casing, so that the nozzles, shaft, runner, hydraulic seal, etc. could be fine-tuned and to ease operation and maintenance work.

Figure 11.12 details the layout of these components inside the powerhouse.

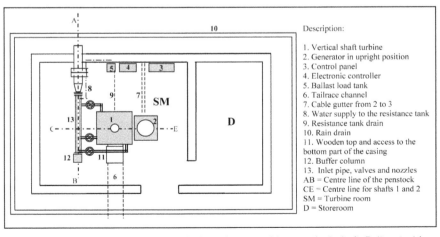

Figure 11.12 Layout of electromechanical equipment with a vertical shaft Pelton turbine, in accordance with the dimensions of the equipment

The distance between the wall and CE depends on the accessories used to join the penstock to the inlet pipe and the turbine nozzles.

The foundation base has the same functions as the base of the horizontal shaft turbine, except that a separate buffer column must be built-in at the free end of the inlet pipe.

Design

The design is similar to the previous case, bearing in mind the axial and radial loads in the turbine and generator shafts.

Axial loads are caused by the weight of the shaft itself and the components attached to it; radial loads are due to the tension of the belts.

These axial and radial loads are absorbed by the upper and lower supports. The upper one is designed to absorb both the axial and radial load and the lower one only the radial load. The pieces that absorb these loads are the bearings duly selected by the expert and installed during the assembly process.

Construction process

The construction process is very similar to that of the foundation for the horizontal shaft Pelton turbine, with a few particular characteristics, as follows.
- Extend the centre line of the penstock and put marks A and B on the floor and M on the wall
- Connect the turbine inlet pipe to the penstock pipe. Then attach the nozzles and valves to the turbine casing and the metal base of the generator (see Figure 11.14 and Photograph 11.2).

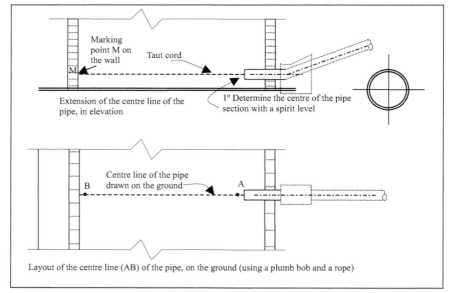

Figure 11.13 Layout of the centre line of the pipe on the ground and on the wall

FOUNDATION FOR THE ELECTROMECHANICAL EQUIPMENT   201

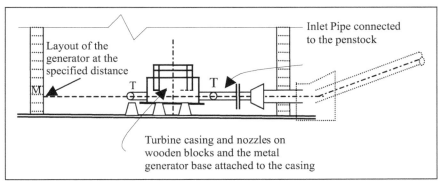

Figure 11.14 Side view of the casing of a vertical shaft turbine and the base of the generator

Photograph 11.2 Pre-assembly of the casing, inlet pipe and alignment of the centre line of the penstock with the centre line of the inlet pipe in order to draw the layout for the excavation of the foundation for the turbine and generator

- Mark line CE on the floor, at right angles with centre line AB. On line CE plot points X_1 and X_2 of the turbine and generator shafts, respectively. On the floor, mark the edges of the casing, the metal base of the generator and other referential points (corners of both the casing and the base of the generator).
- Remove the pre-assembly cords or ropes for the layout.

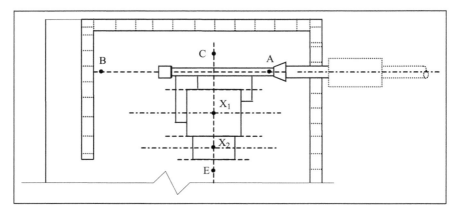

Figure 11.15 Centre line of the penstock (AB), centre line perpendicular to it (CE), and parallel lines that go through the vertices of the casing and the base of the generator

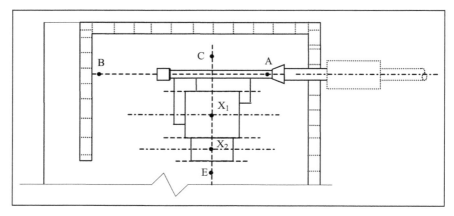

Figure 11.16 Removal of pre-assembled equipment and layout of the casing and the base of the generator on the ground

- Lay out the dimensions for the excavation of the foundation base, in accordance with the positioning of the key points (see Figure 11.17).
- Dig the foundation pit for the turbine and generator to the specified depth. Have a hoist and ropes available to ease the removal of large rocks.
- Shape the foundation pit in accordance with the excavation conditions, as explained in the previous case, unless other characteristics appear, such as water if the pit is close to the river or the water table.
- Verify the foundation depth.

FOUNDATION FOR THE ELECTROMECHANICAL EQUIPMENT

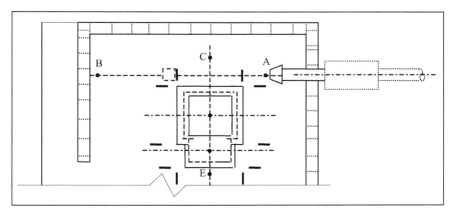

Figure 11.17 Layout of the base for the turbine and the generator and auxilliary components, for excavation purposes

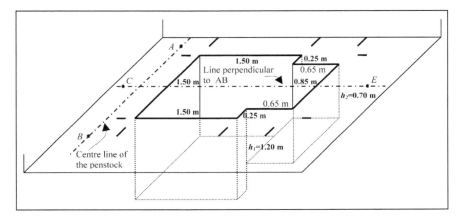

Figure 11.18 Foundation pit

- Compact the soil that will form the base of the foundation floor.
- On the compacted soil, lay a 10 cm thick bed with a C.:Ag 1:10 concrete mix, to obtain a level and uniform foundation surface.
- Draw the centre line of the penstock and the centre of the turbine on the bed, as referential lines for placing the steel reinforcement cage.
- Prepare and fit the reinforcement cage corresponding to the turbine, taking the centre lines drawn during the previous step as a reference. Attach L-shaped steel reinforcement (½" (13 mm) × 0.50 m rebars) and join it to the steel reinforcement cage of the generator base.
- Fix the steel reinforcement cage on concrete blocks, leaving enough space for the minimum coating of concrete (see Figure 11.19).

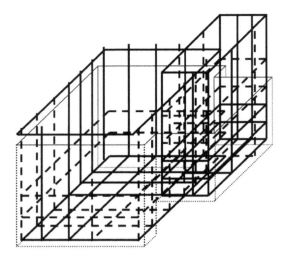

Figure 11.19 Steel reinforcement cage for the foundations of the vertical shaft Pelton turbine and the generator

- Bury the drainpipe of the ballast tank which leads to the turbine pit, leaving the free end of the pipe connected to the formwork, making sure that when the concrete is poured, none gets into the pipe.
- Prepare the concrete mix and pour it on the floor to the required height.
- Again draw the centre lines of the turbine penstock as guidelines for the formwork of the foundation walls.
- Form the foundation walls around the pit, employing a simple method whereby the form can be stripped through the opening to the tailrace channel. Make sure the walls are plumbed and squared to the indicated level and leave an opening of the required width and height for the outlet to the discharge channel.
- On the wall formwork, mark the level to which the concrete should be poured.
- Prepare the concrete mix and pour it on the base of the turbine and generator to the level of the finished floor, applying concrete technology recommendations (mixing, transporting, compacting, etc.).
- Immediately assemble the inlet pipe, valves, nozzles and casing of the turbine and the metal base of the generator, making sure they are level and aligned with the penstock and the perpendicular centre line. The casing is attached to strong wooden planks which rest on wooden blocks on the adjacent floor. Then plumb and fit the anchor bolts onto the casing and the metal base of the generator inside the reinforcement cage, placing the nuts and washers on the free end of the anchor bolts. Immediately complete the formwork on the outside wall of the pit and the generator so that it overlaps the floor and finish pouring the concrete on the base of the turbine and generator (see Figure 11.20).

FOUNDATION FOR THE ELECTROMECHANICAL EQUIPMENT 205

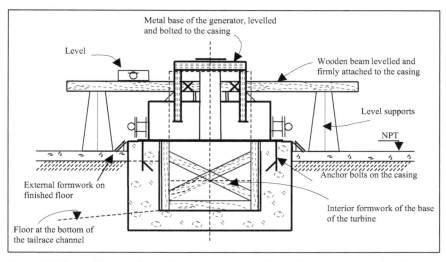

Figure 11.20 Levelling of the casing and the metal base of the generator; forming and cementing the concrete foundation

- Clean the flanges of the anchored casing, the anchor bolts and the screws, removing all traces of concrete.
- Cure the concrete for at least eight continuous days and strip the forms off the walls after a minimum of 10 days.
- Render the floor and walls of the pit and the base of the generator, in accordance with final finish recommendations, remembering to cure the render.

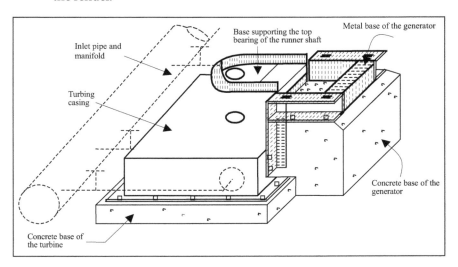

Figure 11.21 Casing of the vertical shaft turbine and metal base of the generator, anchored to their concrete foundations

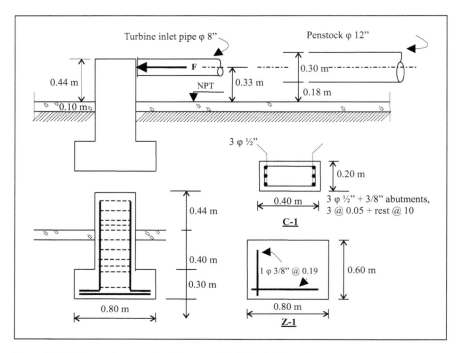

Figure 11.22 Dimensions and reinforcement of the buffer column
Note: 3/8" = 10 mm; ½" = 13 mm; 8" = 20.3 cm; 12" = 30.5 cm

The buffer column is located at the free end of the turbine inlet pipe, embedded in the floor to withstand the thrust of the water. The column is designed in the same way as a small concrete cantilever beam with a point load.

Technical specifications

These are very similar to the technical specifications for the foundation of the previous case.

- Concrete f'_c =175 kg/cm² (proportions by volume → cement:sand: pebbles → 1:2:3)
- Steel: f_y = 4,200 kg-f/cm²
- Casing of the vertical and horizontal steel: at least 2.5 cm
- Rendering of inside and outside walls with 3:1 sand-cement mortar, with polished cement.
- Fix anchor bolts on the casing, at least 0.30 m, and apply at least 0.07 m of coating around them.
- Fix the anchor base of the generator on the turbine casing, then anchor it to the concrete base.
- σ_T = 1.5 kg-f/cm² (verify on site).
- Outlet of the tailrace channel: 0.60 wide × 0.75 high.

- Other specifications that the project designer may consider worth bearing in mind when building the foundation bases.

Foundation for the Michell-Banki turbine

The authors have been involved in the installation of several Michell-Banki (cross-flow) turbines below 100kW. Relatively small foundations were required in these cases.

The layout of the electromechanical equipment inside the powerhouse and the technical criteria employed are very similar to the horizontal shaft Pelton turbine because Michell-Banki turbines also have a horizontal shaft with a transmission system of pulleys and belts. The same criteria should be applied for installations with direct coupling, but bearing in mind the small differences due to the absence of pulleys and belts.

Figure 11.24 shows a typical layout of electromechanical equipment (elements 5, 8 and 9 are not required if the ballast load is air-cooled, which is customary for less than 5 kW of power).

The function of the foundation is the same as for the horizontal shaft Pelton turbine.

Design (Figure 11.24)

Follow the same procedure as for the horizontal shaft Pelton turbine. Access underneath the casing is not usually required; therefore, there is no need for the pit at the base of the turbine (which also forms the tailrace channel) to be any larger than normal.

The sizing of the inside of the pit will depend on the measurements of the metal base of the turbine and generator. It is a single base shared by both units, on which the casing and the generator are bolted down. Likewise, the depth

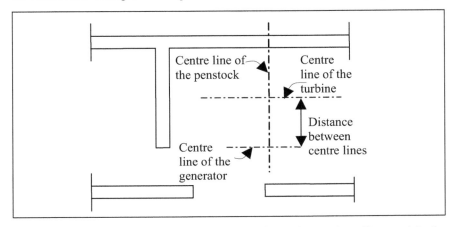

Figure 11.23 Location of the centre lines of the turbine and generator with respect to the centre line of the penstock

208 DESIGNING MINI AND MICRO HYDROPOWER SCHEMES

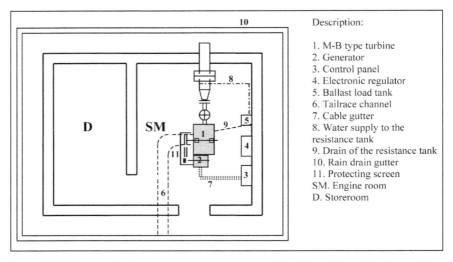

Figure 11.24 Layout of electromechanical equipment with a Michell-Banki turbine, similar to that of the horizontal shaft Pelton turbine

includes the breadth of the floor, the height of the opening to the tailrace channel and a short plinth above the level of the finished floor.

The total static load, calculated in kg-f, is the same as the calculation for horizontal shaft turbines which, divided between the horizontal area in contact with the base in cm², should be less than the bearing capacity of the foundation soil.

Construction process

Repeat the preliminary steps for the layout, excavation, preparation and assembly of the steel reinforcement cage. Then lay the paving at the bottom of the pit and form the walls and the outside of the base for the generator and the turbine.

When placing, levelling, aligning and fastening the metal base to be anchored, make sure that, in addition to being level and aligned, there is a 1 to 2 cm separation between the turbine inlet pipe and the threaded nipple that joins the valve connected to the penstock. The separation between the nipple and the turbine inlet is subsequently covered with a dresser union (see Figure 11.25).

Technical specifications

These are the same as for the quality of the concrete and steel used:
- Concrete f'_c = 175 kg/cm² (volume dose → cement:sand:stones → 1:2:3)
- Steel: f_y = 4,200 kg-f/cm²
- Vertical and horizontal steel coating: at least 2.5 cm
- Inside walls plastered with 3:1 sand-cement mortar, polished cement finish.

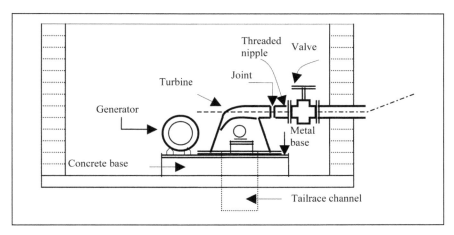

Figure 11.25 Levelling, alignment and separation joint between the Michell-Banki turbine and the nipple joined to the valve, before pouring the concrete

- Specify the dimensions and breadth of the metal base and the minimum depth of the anchor.
- $\sigma_{T\,max} = 1.00$ kg-f/cm^2 (verify on site).
- Width and height of the outlet window of the discharge channel.
- Other technical specifications considered by the turbine manufacturer as worth bearing in mind when building the foundation base.

Foundation for the axial turbine with a vertical shaft

To explain in a practical way, an example of the installation of a propeller turbine at Las Juntas, Peru, is described. This was the first turbine that Practical Action manufactured and installed in Peru (1999).

$Q = 0.600$ m^3/s
$H_B = 7.25$ m
$P = 25$ kW (electrical power)

The turbine consists of a spiral shaped casing with a circular section of a varying diameter.

The guide vanes, shaft and propeller-shaped runner are located in the middle and the draft tube that conveys the water to the tailrace channel is located underneath the runner.

The draft tube has three parts. The first part is a vertical cone; the second part is referred to by some authors as a knee or elbow pipe, which changes the direction of the water outlet, forming a bend, with a variable sequential geometric section; the third part is an outlet section, which has an increasing cross-sectional area.

For the layout of the electromechanical equipment, the technical criteria applied are similar to those of the vertical shaft Pelton turbine, except that it

210 DESIGNING MINI AND MICRO HYDROPOWER SCHEMES

does not have a turbine manifold. The penstock is connected directly to the valve and the valve to the casing with a dresser union.

The casing is a spiral pipe with an initial diameter the same as the diameter of the penstock (600 mm), which decreases in size until it reaches the aperture of the guide vanes. Steel brackets are welded to the outer circumference of the spiral in a vertical position, to be attached and bolted down on the metal base of the generator, in a similar way to the vertical shaft Pelton turbine. These welded brackets have two supports at the bottom held down by anchor bolts, as well as two lower supports under the spiral casing.

In the Las Juntas MHS, the layout of the electromechanical equipment was as follows, in accordance with the dimensions and design established by Practical Action (see Figure 11.26).

Function of the turbine foundation. Confirm the design height, bearing in mind that the height is measured from the water surface in the forebay tank to the water surface in the tailrace channel. It is therefore important to determine the turbine foundation level and the lower level of the draft tube and to fix the level of the water surface in the tailrace channel.

Fix the turbine and generator on the powerhouse floor so that the generator can deliver the estimated electricity with the mechanical energy created by the turbine.

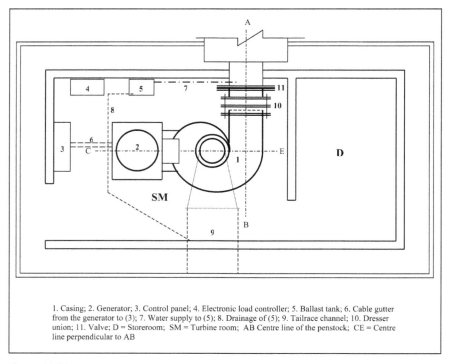

1. Casing; 2. Generator; 3. Control panel; 4. Electronic load controller; 5. Ballast tank; 6. Cable gutter from the generator to (3); 7. Water supply to (5); 8. Drainage of (5); 9. Tailrace channel; 10. Dresser union; 11. Valve; D = Storeroom; SM = Turbine room; AB Centre line of the penstock; CE = Centre line perpendicular to AB

Figure 11.26 Layout of electromechanical equipment with a low head turbine (Kaplan propellor type), 'Las Juntas' MHS, Peru

Design

Sizing of the foundations

The foundation of the axial turbine comprises three small concrete bases isolated from each other. The three supports on the bottom of the casing are anchored to these bases with anchor bolts.

There is no need to dig a pit underneath the turbine to receive the water from the turbine, as this is replaced by the draft tube which is directly connected to the tailrace channel. The plan view sizing of these bases is determined with the dimensions of the casing supports, adding the necessary width, length and depth to these measurements, so that the anchor bolts are completely embedded in the concrete.

The sizing of the generator base is obtained with the measurements of the metal base attached and bolted to the turbine casing, also adding the necessary measurements to ensure that the anchor elements are completely covered in concrete.

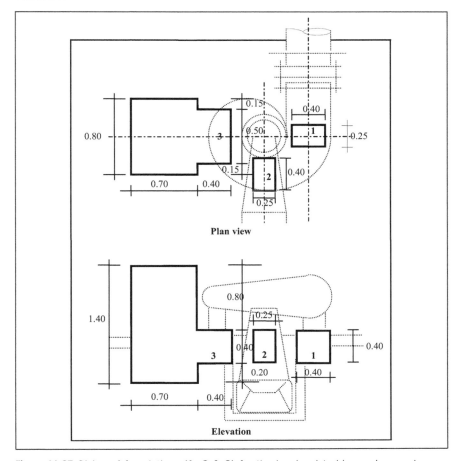

Figure 11.27 Sizing of foundations (1, 2 & 3) for the low head turbine and generator

212 DESIGNING MINI AND MICRO HYDROPOWER SCHEMES

The depth of the generator is measured from the foundation level to the level of the anchor for the metal base.

Calculation of the total static load

1. Weight of the base:
   = (2 × 0.4 × 0.4 × 0.25 + 0.7 × 0.80 × 1.40 + 0.4 × 0.4 × 0.5)2,400 kg-f/m³
   = 2,316 kg-f
2. Weight of the water inside the casing: (approximately):

   $= \frac{\pi}{4} \times D_i^2 \, L \times \gamma_a = \frac{3.14}{4} \times 0.40^2 \times 3.14 \times 1.30 \times 1000$

   kg-f/m³ = 512.70 kg-f
   D_i = average diameter
   L = average length of the casing
3. Weight of the equipment
   Piping and accessories = 507 kg-f
   Generator and accessories = 255 kg-f
4. Additional weight from rotating elements:
   = 12(Shaft + Pulley + Runner) + 12 × 180 × 0.50 = 2,004 kg-f
5. Live load = (0.70 × 0.8 + 0.40 × 0.50 + 2 × 0.25 × 0.40)250 kg-f/m²
   = 240 kg-f

   Total static load = 2265.60 + 512.70 + 507 + 255 + 2004 + 240 = 5,584.30 kg-f

Table 11.1 Characteristics of the equipment supplied to 'Las Juntas' MHS

Equipment	Supplier	Element	Weight (kg-f)	Dimensions
Low head turbine (with ITDG technology)	Tepersac (Peru)	Turbine inlet pipe, elbows	—-	
		Valves (01)	80	700 × 700 mm card type
		Casing	210	
		Shaft	35	Φ = 60 mm × 750 mm
		Pulley	22	
		Runner	20	
		Supports, bearings, runners, guide vanes	50	
		Dresser union	90	600 mm
Three-phase asynchronous generator, self-excited, 220V, 60 Hz 1,800 RPM, 25 kW	Stamford	Unit, including pulley U-shaped anchor base and support attached to the casing. Base on anchor	180 55 20	Φ = 40 mm × 600 mm

Bearing capacity

Bearing capacity = 5584.30 kg-f/0.96 × 10,000 cm² = 0.58 kg-f/cm²

Verification:

$\sigma_{T\ admissible}$ > Bearing capacity
2 kg/cm² > 0.58 kg-f/cm² ... OK

Steel layout (for reinforcement)

See Figure 11.28.

Construction process

- Level the natural ground of the work site.
- Extend the centre line of the penstock and mark the points on both walls, e.g. A and B (see Figure 11.29).
- Connect the casing to the penstock, fit the supports and level.
- On the ground, mark the straight line PQ, parallel to the centre line of the penstock (AB) and line (RS) at a right angle with PQ and AB. Lines RS and PQ meet in the middle of the casing at point O (see Figure 11.30).
- Remove the casing and mark the site of the draft tube on the ground, taking its top measurements.
- Dig the ditch for the draft tube to the design depth.
- Remove the dug-out material and place the draft tube in the ditch at the specified level.

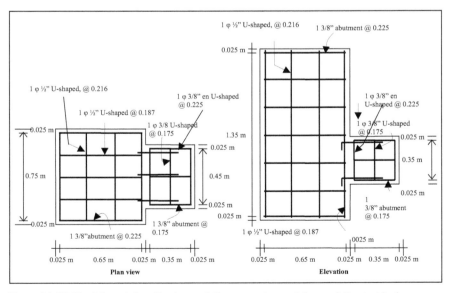

Figure 11.28 Steel layout (on the base of the generator and base of the turbine)

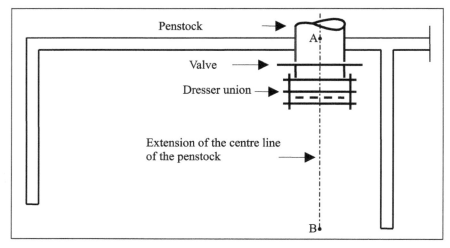

Figure 11.29 Layout of the centre line of the penstock

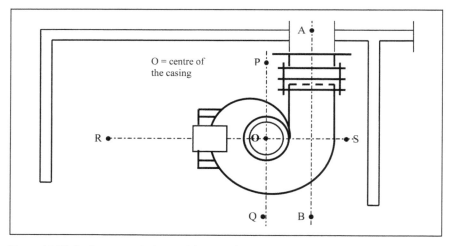

Figure 11.30 Casing presentation and layout of centre lines

- Build the floor at the bottom of the ditch and the U-shaped walls to the specified level with reinforced concrete, to support and protect the sides of the draft tube (see Figures 11.31 and 11.32).
- Pre-assemble the casing with the draft tube and the metal base of the generator.
- Connect this unit to the penstock, making sure it is aligned with the centre lines.
- Level all the assembled units, taking referential levels for the finished floor and the points defining the line at a right angle with the centre

FOUNDATION FOR THE ELECTROMECHANICAL EQUIPMENT 215

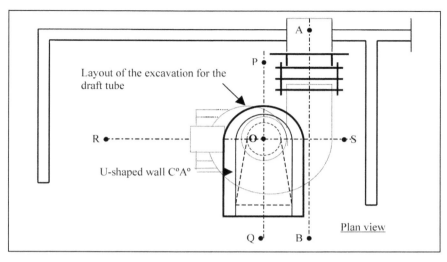

Figure 11.31 Layout of the protecting wall for the draft tube

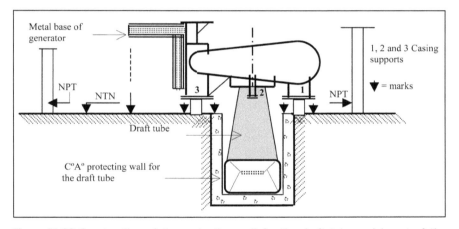

Figure 11.32 Construction of the protecting wall for the draft tube and layout of the foundations

line of the penstock, on which the centre lines of the turbine and generator will be drawn.
- Draw the positions of the casing supports containing the anchor bolts of the turbine and the generator base.
- Remove the casing with the metal base of the generator and the draft tube.
- Draw the measurements of the bases for excavation purposes.
- Dig the foundation for the generator to the specified depth (see Figure 11.33).

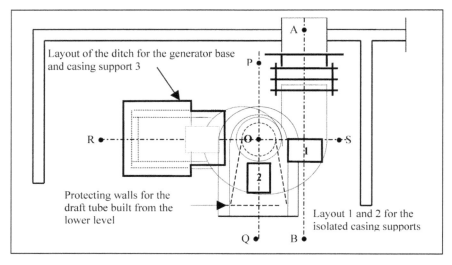

Figure 11.33 Layout of the ditch for the generator base, walls for the draft tube and casing supports

- Build a 10 cm cement and reinforced concrete bed in the foundation for the generator, making sure it is levelled.
- On this bed, draw the line at a right angle to the penstock to guide the installation of the steel reinforcement cage and formwork.
- Place the steel reinforcement cage on the base of the generator.
- Prepare the formwork for the outside walls of the base for the generator and the casing.
- Pour the first layer of the base to the top level of the excavated ditch.
- Connect the turbine casing to the penstock, bolting the casing onto the metal base of the generator and the draft tube, then fix, level and align them to finish laying the concrete.
- Form the outside walls of the generator base, ensuring that they are vertical and then fasten the anchor bolts on the metal base of the generator.
- Prepare the concrete mix and pour it in the formwork, applying the recommendations to obtain good quality concrete.
- Strip the forms after 24 hours but do not remove the supports holding the weight of the casing and the metal base of the generator until the concrete has set for at least 10 days (see Photograph 11.3).
- Render the outside walls with 3:1 sand-cement mortar, polished finish.
- Bases 1, 2 and 3 of the casing anchor are built separately from the generator base. Supports 1 and 3 are built when the protecting wall is finished and support 2 after filling in and compacting the ditch for the distribution pipe. The steel reinforcement for base 1 fits into the steel of the protecting wall of the draft tube. Base 3 goes with the steel of the generator base or the wall. Base 2 has no steel reinforcement (see Figure 11.34).

FOUNDATION FOR THE ELECTROMECHANICAL EQUIPMENT 217

Photograph 11.3 Metal base for the generator, anchored to its concrete foundation. The turbine casing is coupled to the draft tube and the metal base of the generator

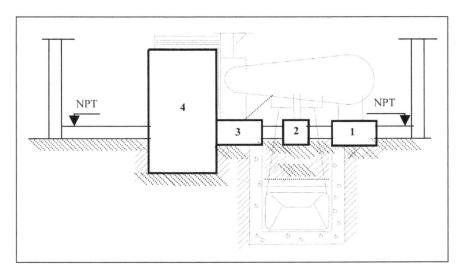

Figure 11.34 Foundations for the turbine and generator
Note: 1, 2 and 3 = turbine foundation; 4 = generator foundation

Technical specifications

- Concrete f'_c = 175 kg/cm² (volume dose → cement:sand:stone → 1:2:3)
- Steel: f_y = 4,200 kg-f/cm²
- Covering of the vertical and horizontal steel: at least 2.5 cm.
- Render the outside walls with 3:1 sand-cement mortar, polished cement finish.
- Specify the dimensions and breadth of the metal base and the minimum depth for the anchor.
- $\sigma_{T\ minimum}$ = 2.00 kg-f/cm² (verify on site).
- Other technical specifications recommended by the manufacturer of the turbine and generator.

CHAPTER 12
Assembly of the turbine and generator

This chapter covers the steps required to put together the different types of turbine dealt with in the chapter on the foundation. It is a task that consists of assembling the turbine elements with care and precision, before the turbine is finally anchored to the foundation. The generator is then fixed onto its respective foundation.

The end result is to ensure that the turbine and generator work properly, coupling them with the elements that constitute the transmission system to efficiently obtain the power estimated in the design.

Assembly of the horizontal shaft Pelton turbine with two jets

Turbine components

Before installing the turbine components, assemble the accessories required to connect the penstock to the turbine. These are (Figure 12.1):
- FU PVC reducer;
- RU PVC flange union;
- dresser union (optional);
- others (depending on the case).

The turbine components are as follows (Figure 12.2):
1. inlet pipe;
2. valves;
3. elbows;
4. nipple;
5. gaskets;
6. casing (bottom and top base);
7. nozzles;
8. shaft;
9. bearing support base;

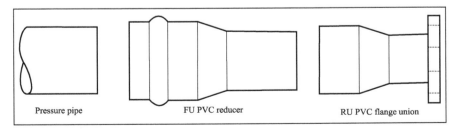

Figure 12.1 FU and RU PVC accessories

220 DESIGNING MINI AND MICRO HYDROPOWER SCHEMES

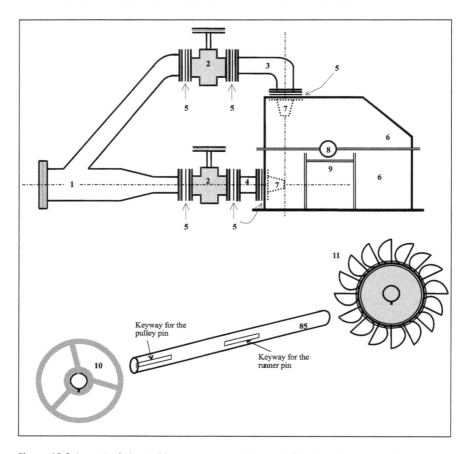

Figure 12.2 Layout of the turbine components (see text for list of components)

10. pulley and taper fixing;
11. runner;
12. other elements: bearings, hydraulic seals, belts, bolts, nuts, washers, locking screws, etc.

Assembly of the turbine

The following are referential guidelines for the installation of horizontal shaft Pelton turbines:
1. Install the turbine inlet pipe in the penstock, using the reducer and the flange union (expansion coupling may or may not be necessary).

ASSEMBLY OF THE TURBINE AND GENERATOR

The flexible union PVC reducer is joined to the penstock with the rubber ring, to which end only lubricant is required. The RU PVC flange union is welded to the reducer with special glue for PVC material.

The reducer and the flange union are of equal pressure rating to the penstock, if not higher.

A gasket is placed between the inlet pipe and the flange union. Both are joined with bolts, nuts and washers (see Figure 12.3).

For steps 2 to 7, see Figure 12.4.

2. Connect the valve to the bottom side of the inlet pipe, using bolts, nuts, washers and rubber and canvas gaskets (apply a film of grease or suet to both sides of the gaskets to prevent them from sticking to the contact surfaces and to ease subsequent maintenance activities).
3. Couple the nipple to the valve of the bottom nozzle, then to the casing. In both cases, use bolts, washers, nuts and gaskets.
4. Clean the metal base of the turbine already anchored to the concrete base, as well as the contact surfaces of the lower base of the casing.
5. To ease the assembly, it is advisable to apply a layer of silicone around the edge of the anchored metal base instead of a gasket. The silicone layer must be of a standard width and thickness, leaving the holes free for the bolts (see Photograph 12.1).
6. On the anchored base, join the lower base of the casing with a layer of silicone, then put in the bolts, washers and nuts all around it. Do not tighten the bolts permanently yet.
7. Install the bottom nozzle, using the selected bolts, nuts and washers.

For steps 8 to 18, see Figure 12.5

8. Install the runner on the shaft, using the wedge key.
9. Place the runner with the shaft and the hydraulic seals on the casing to try out their position and the space that should be left free between the ends of the blades and the outlet of the bottom nozzle.

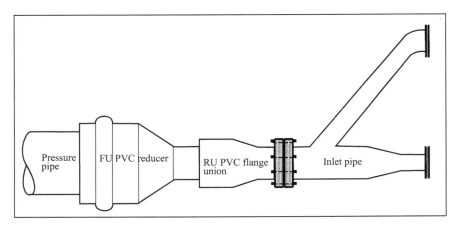

Figure 12.3 Assembly of the turbine components

222 DESIGNING MINI AND MICRO HYDROPOWER SCHEMES

Photograph 12.1 Application of the silicone seal

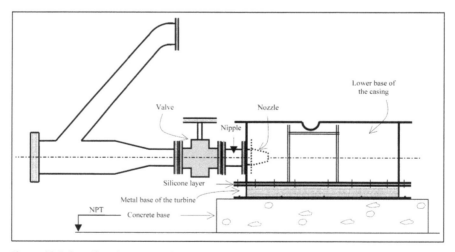

Figure 12.4 Coupling the valve, nipple, casing and nozzle on the metal base

10. Put the two plumber blocks on the supporting base of the casing and install the two roller bearings that encircle the shaft (see Photograph 12.2).
11. Permanently centre the runner with the bottom nozzle and adjust the roller bearings, plumber blocks, nozzle, casing, etc. with the bolts, nuts and washers of each component.

ASSEMBLY OF THE TURBINE AND GENERATOR

Photograph 12.2 Fitting the plumber blocks and roller bearings

12. Couple the top nozzle to the top part of the casing.
13. Clean and apply a coat of silicone to the edge of the bottom part of the casing in contact with the top part, forming a gasket of a uniform width and breadth, leaving the holes free for the bolts.
14. Attach the top part of the casing to the bottom part, using bolts, washers and nuts.
15. Install the second nozzle valve of the top part with the elbow and the inlet pipe, putting in the gaskets, bolts, washers and nuts.
16. Finally tighten the bolts, verifying the level and alignment. As a precaution, always keep the valves shut. At no time during the assembly of the turbine should the penstock contain water.
17. Adjust the seals, removing and then replacing the metal top on the top part of the casing.
18. Put the turbine in the specified position at the free end of the shaft, fixing it with the tightening key and taper bush.

Assembly of the generator (Figure 12.6)

1. Place the generator on the metal base anchored to the foundation base, with the help of a hoist, and put in the bolts, washers and nuts without tightening them, trying to maintain the specified distance between the shafts of the generator and the turbine.
2. Attach the generator pulley to the free end of the shaft with the wedge key and locking screws.

224 DESIGNING MINI AND MICRO HYDROPOWER SCHEMES

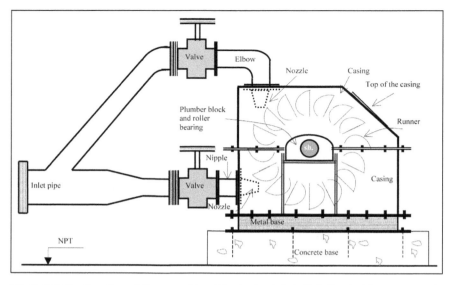

Figure 12.5 Continuation of the assembly of the horizontal shaft Pelton turbine components

3. Align the generator pulley with the turbine pulley, using the vertical section of both pulleys as a reference and using a thin, taut cord. Then tighten the bolts on the base of the generator.
4. Place the transmission belts on the pulleys and tighten them as necessary with the screw adjusters system fixed on the metal base of the generator. In practice, the belts are considered taut when no more than a 1" (2.5 cm) sag appears when the belts are pressed down in the middle with a finger.

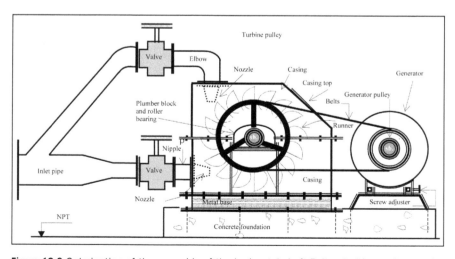

Figure 12.6 Culmination of the assembly of the horizontal shaft Pelton turbine and generator

Assembly of the vertical shaft Pelton turbine with three jets

The accessories previously installed in the penstock can be the same as for the previous case. In other cases, they may vary due to the larger diameter of the penstock and/or inlet pipe, or because of their cost.

Figure 12.7 shows the components of this turbine. Some elements vary in quantity, such as elbow pipes, valves, gaskets, bolts, nuts, etc., depending on the number of nozzles. The casing is usually a single piece.

Assembly of the turbine

This consists of two stages:

Stage one

Stage one (see Figure 12.8) comprises the assembly of the accessories to be installed in the penstock, then the coupling of the inlet pipe to the valves,

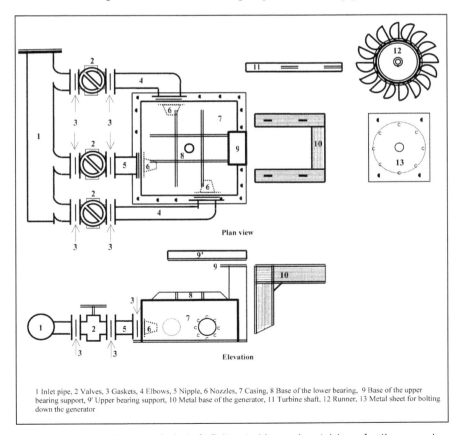

1 Inlet pipe, 2 Valves, 3 Gaskets, 4 Elbows, 5 Nipple, 6 Nozzles, 7 Casing, 8 Base of the lower bearing, 9 Base of the upper bearing support, 9' Upper bearing support, 10 Metal base of the generator, 11 Turbine shaft, 12 Runner, 13 Metal sheet for bolting down the generator

Figure 12.7 Pieces of the vertical shaft Pelton turbine and metal base for the generator

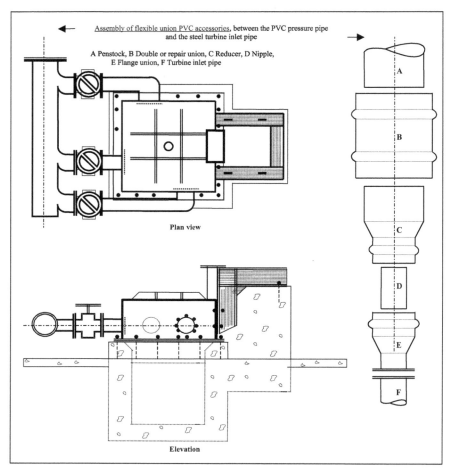

Figure 12.8 First stage of the assembly of a vertical shaft Pelton turbine and metal generator base

elbows, nipple and the casing. Finally, the metal base of the generator is attached to its casing.

All these pieces should be assembled during the construction of the foundations for the turbine and generator. At this stage, it is also important to ensure the precision of the levelling and alignment required for the assembly of the other components.

Stage two

Stage two consists of the assembly of all the other components, in accordance with the following guidelines (see Figure 12.9).

1. Put the nozzles in place with the bolts, washers, nuts and gaskets.
2. Attach the shaft to the runner, fixing it with the wedge key and placing the disk and security bolt on the runner at the bottom end of the shaft.
3. After the runner, put seal elements in place, to be fitted later.
4. Through the tailrace channel under the casing, gain access to the hole in the middle of the casing and, working upwards, attach the components assembled in accordance with steps 2 and 3 to the free end of the shaft. Keep this group of parts suspended with the help of a hoist.
5. Centre the runner with the nozzles, so that the horizontal section determined by the runner coincides with the horizontal section determined by the three nozzles, making sure that the mouths of the nozzles do not brush against the runner buckets.
6. Make sure the runner is centred, placing it on its respective bearing and support on the bottom.
7. Place the pulley and its taper bush on the shaft.
8. Tighten the pulley on the shaft at the corresponding level, using the wedge key and the taper bush.
9. Couple the bearing support to the top support, using the corresponding bolts, washers and nuts.
10. Assemble the base and the top roller bearing, making sure the shaft is plumb-lined (vertically).
11. Firmly tighten all the bolts, checking the level of the runner and pulley and the upright position of the shaft.
12. The safety screen should be fitted after the belts have been placed on the pulleys of both the turbine and generator.

Assembly of the generator

The following are guidelines for a typical scheme.
1. Attach the metal plate to the generator with the corresponding bolts.
2. Place the pulley on the free end of the generator shaft, fixing it in the right position with the wedge key and locking screw. This pulley should be aligned with the turbine pulley.
3. Place all the parts assembled in the previous steps on the metal base anchored to the concrete base, with the help of a hoist, bolting it down with bolts, nuts and washers.
4. Check the alignment of the turbine and generator pulleys, to prevent undesirable vibrations during their operation.
5. Put the belts on both pulleys and tighten them with the screw adjusters on the generator base.
6. Tighten all the adjusting nuts on all assembled parts.
7. Put the safety screen in place.

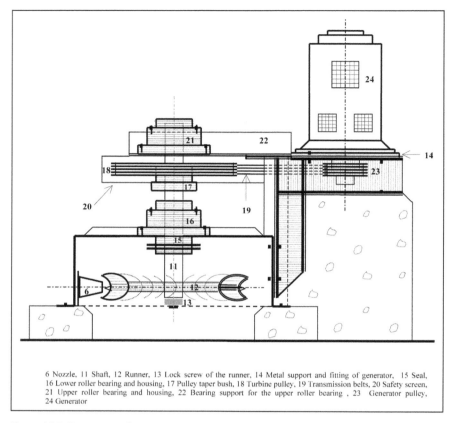

6 Nozzle, 11 Shaft, 12 Runner, 13 Lock screw of the runner, 14 Metal support and fitting of generator, 15 Seal, 16 Lower roller bearing and housing, 17 Pulley taper bush, 18 Turbine pulley, 19 Transmission belts, 20 Safety screen, 21 Upper roller bearing and housing, 22 Bearing support for the upper roller bearing , 23 Generator pulley, 24 Generator

Figure 12.9 Stage two of the assembly of the vertical shaft Pelton turbine and generator
Note: part numbering is continued from Figure 12.7

Assembly of the Michell-Banki turbine (Figure 12.10)

Small turbines of 4, 7, 11, 14 and 40 kW of power have been installed in the past. On those occasions, the components of these turbines were assembled by the manufacturer in the workshop and then transported to the site in a single unit, already coupled to the metal base ready for cementing.

The parts to be assembled on the site are the components of the transmission system (pulleys and belts) and the accessories required to attach them to the penstock (dresser union, nipple, valve and FU PVC flange union).

The elements composing a Michell-Banki turbine are basically the casing, the support base for the plumber blocks and the inlet pipe (which form a single piece with the casing), the shaft, runner, runner inlet valve, guide vane, plumber blocks and bearings.

The generator is assembled on the metal base. The holes in the base indicate the accurate position. It is then aligned, taking the outer and inner sides of the

ASSEMBLY OF THE TURBINE AND GENERATOR 229

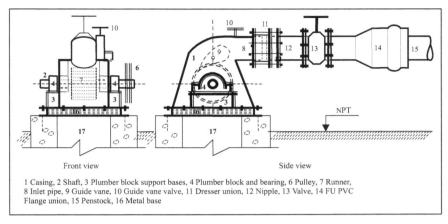

Figure 12.10 Assembly of the Michell-Banki turbine and accessories for coupling it to the penstock

1 Casing, 2 Shaft, 3 Plumber block support bases, 4 Plumber block and bearing, 6 Pulley, 7 Runner, 8 Inlet pipe, 9 Guide vane, 10 Guide vane valve, 11 Dresser union, 12 Nipple, 13 Valve, 14 FU PVC Flange union, 15 Penstock, 16 Metal base

turbine and generator pulleys as a reference. The belts are then tightened properly with the screw adjusters.

In the case of more powerful Michell-Banki turbines, the components are taken apart and transported to the site to be assembled in a similar way to the horizontal shaft Pelton turbine.

Photograph 12.3 Michell-Banki turbine and generator

Assembly of the axial turbine and generator

Turbine assembly

This involves two stages

Stage one (Figure 12.11)

Stage one consists of installing part of the turbine components and the metal base of the generator during the construction of the foundation.

The following are the components to be installed and the steps to be followed.
1. Couple the casing to the penstock with the valve and the expansion coupling.
2. Join the draft tube to the casing, using the bolts, washers, nuts and gaskets (or silicone).
3. Assemble the metal base of the generator, with the turbine casing.
4. Make sure that the turbine casing and the metal base of the generator are aligned, levelled and anchored to their respective foundation.

Stage two (Figure 12.12)

1. Place the inner guide vane ring in the central cavity of the casing, using the corresponding bolts, washers and nuts.
2. Assemble the shaft with the runner, using the wedge key.
3. Fix the nose cone with a screw on one end of the shaft on the bottom side of the runner (see Photograph 12.4).

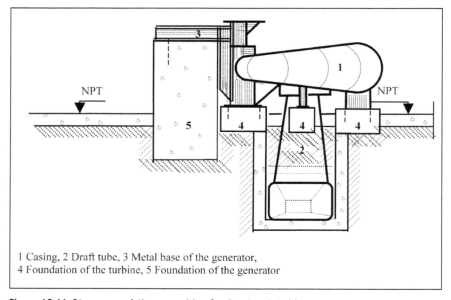

1 Casing, 2 Draft tube, 3 Metal base of the generator,
4 Foundation of the turbine, 5 Foundation of the generator

Figure 12.11 Stage one of the assembly of a low head turbine

ASSEMBLY OF THE TURBINE AND GENERATOR

Photograph 12.4 Axial turbine, showing the nose cone, runner, diffuser section and the lower bearing housing, installed together on the shaft

4. Place the diffuser section and the lower bearing housing on the turbine shaft (see Photograph 12.4).
5. Attach the elements fitted in the previous step, using the corresponding bolts, washers and nuts.
6. Assemble the lower bearing and put the packer in place.
7. In the central cavity of the casing, couple the assembled elements described in steps 2 to 6, bearing in mind the level of each one with respect to the central guide vane ring and the upright position of the shaft.
8. Place the pulley on the shaft and fix it with the wedge key and taper bush, at the corresponding level.
9. Couple the upper bearing support to the casing support structure.
10. Assemble the upper bearing and check that the shaft is truly vertical.
11. Tighten the bolts of the different assembled elements to begin the operating tests.

Assembly of the generator

This procedure is similar to the assembly of the generator for the vertical shaft turbine.
1. Attach the metal plate to the generator with the corresponding bolts.
2. Place the pulley on the free end of the generator shaft, fitting it in the specified position with the wedge key and locking screw. This pulley must be aligned with the pulley of the turbine.
3. Place all the parts assembled in the previous steps on the metal base anchored to the concrete base, with the help of a hoist and bolting it down with the bolts, nuts and washers.
4. Verify that the pulleys of the turbine and generator are aligned with precision.
5. Place the belts on both pulleys and check their tension, using the screw adjusters on the base of the generator.
6. Tighten all the nuts on the assembled pieces.
7. Put the safety screen in place.

Cable connections

From the generator to the control panel

The generator and the control panel are connected via cables, depending on the generating power and the existing connection system. For example, if the generator has a star 380/220 V (or 415/240 V) three-phase connection, the connection to the control panel will be via four cables – three live phases and one neutral phase. If it is a delta 220 V or 240 V three-phase connection, the connection to the control panel will be via three cables.

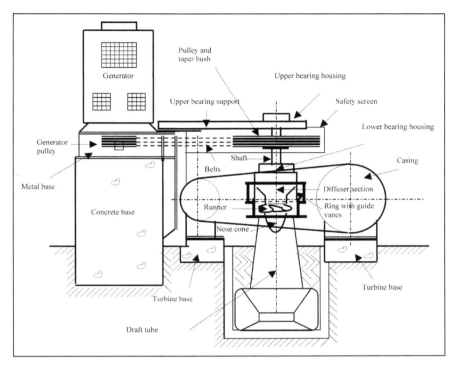

Figure 12.12 Final assembly of a low head turbine and generator

The rating of the cables is selected in accordance with the maximum amperage they will bear, based on the generating power.

From the control panel to the electronic load controller

The design of the electronic load controller is based on the generating load, the objective being to dissipate all the power generated when the main load is not supplied; therefore the cables used for the connections should be of the same rating as the connection between the generator and the control panel. The number of cables depends on the connection system of the controller (three or four lines).

From the electronic load controller to the ballast load

The connection between the controller and the ballast load depends on the number of elements contained in the ballast load and the arrangements made, in accordance with the energy to be dissipated, which could be all or part of the power generated. In some cases the electronic controller has a separate connection to each element (resistance).

Earthing of the electromechanical equipment

The electromechanical components that require earthing are the generator, the control panel and the electronic load controller. The earth connection is a bare copper cable leading to a copper rod covered with water soluble salts and fertile earth, in accordance with electricity standards.

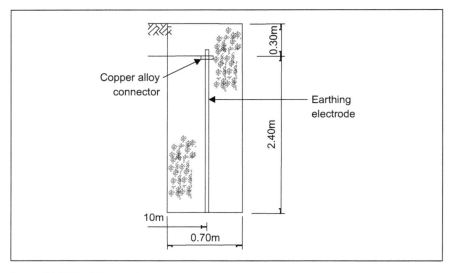

Figure 12.13 Earthing system

CHAPTER 13
Tailrace channel

The tailrace channel begins underneath the turbine and carries the water from the turbine to the river.

In some cases the water is not carried directly to the river but is used for irrigation or in fish farms.

The tailrace channel is located at the end of all the civil works, usually from the powerhouse to the river.

It may be a short channel or several dozen metres long. Its positioning bears a relation to the initial level and the final outflow level. The initial level is the bottom of the turbine foundation and the final level corresponds to the point where it reaches the river bank. This final level must be higher than the maximum rainy season level of the river (see Figure 13.1).

In the plant, the tailrace channel exits through one side of the turbine foundation pit, across the foundation of one of the walls of the powerhouse and when it reaches the river it should form a 45° angle with the centre line of the river. The connecting cables between the generator and the control panel must be prevented from crossing the tailrace channel.

Design of the tailrace channel

The same technical criteria used for the design of the conveyance channel are applied, bearing in mind the type of material selected. Open channels are usually

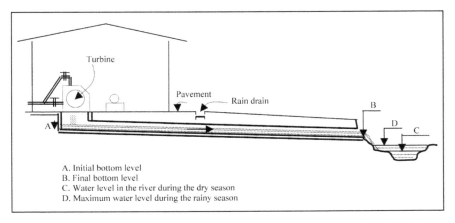

Figure 13.1 Location of the tailrace channel (elevation) for Pelton or Michell-Banki type turbines

made of concrete; PVC pipes are used in some cases when permitted by the conditions of the soil, flow, slope, etc.

The section of this channel can be the same as the section of the conveyance channel, although this may be reduced by increasing the slope and the speed to permissible values.

The design of the tailrace channel is the same for horizontal or vertical shaft Pelton turbines and for Michell Banki turbines, providing the flow is the same, with certain characteristics recommended by the turbine manufacturer to ease some of the assembly or maintenance operations. For example, for Pelton turbines with a vertical shaft, it is recommended that the design should include an access underneath the casing, for assembly and maintenance.

Special characteristics are required for the design of the tailrace channel for reaction turbines (which differ from impulse turbines such as the Pelton and Michell-Banki). At the beginning of the channel the water is not freely discharged, but passes through a draft tube (a diverging cone). In the middle part of the channel a concrete dyke raises the water level to ensure flooded suction conditions. These two characteristics should be previously coordinated with the turbine designer, so that the designed channel responds to the working conditions of the turbine (see Figure 13.2).

Construction process

Tailrace channels for Pelton or Michell-Banki turbines

- Draw the centre line of the channel from the bottom of the turbine base to the river shore, in the direction considered in the design.
- Check the level at the beginning and at the end of the channel on the river shore, bearing in mind that the length, slope and outflow level must be a good deal higher than the maximum water level during the rainy season.

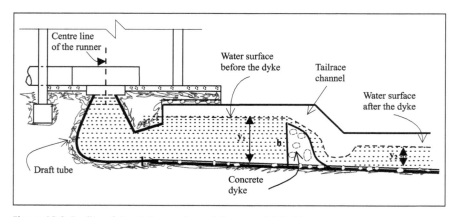

Figure 13.2 Profile of the tailrace channel for an axial turbine

- Supervise and make sure that, when laying the foundation for the turbine, the above-mentioned levels are fixed and that permanent formwork is fitted in one of the walls of the turbine foundation, as a window or outlet for the water from the turbine.
- Once the foundation of the turbine has been poured and the concrete is set, the temporary formwork is removed and the initial level is checked again.
- Taking the central line of the channel as a reference, draw the total width for excavation purposes.
- Dig the channel to the specified depth, in accordance with the design slope, and remove surplus material.
- Shape the profile of the channel ditch and put level rods in place in the bottom of the channel every 5 m, taking the design slope as a reference.
- Pour a layer of concrete of at least $f'_c = 140$ kg/cm² and between 12 and 15 cm thick. If the channel is not made of concrete, put in a channel bed and install the PVC piping considered in the design.
- Form the channel walls, prepare the right proportion of materials for the concrete and pour it in up to the design level. For safety reasons, put an iron fence in a vertical position at the crossing of the channel and the foundation of the wall.
- Inside the powerhouse, the tailrace channel is covered with the concrete that forms part of the floor of the powerhouse and another part has a wooden cover flush with the floor, which allows access to the area underneath the casing if it is a Pelton turbine with a vertical shaft.
- Render over the floor and walls of the tailrace channel. If the channel is a PVC pipe, cover the ditch and add at least one manhole for maintenance tasks.
- Curing the concrete and the rendering is very important for the useful life of the tailrace channel.
- Reinforce the ground at the end of the tailrace channel by the river with concrete, to avoid its erosion by the river current. This is usually done with stone masonry, bedded with sand and cement mortar.

Tailrace channel for a reaction turbine

The procedure is similar to the previous case. Here, the initial level is the bottom of the draft tube, the channel is deeper, the excavation work, formwork, laying the concrete and coating, etc. are also more burdensome and, in addition, a dyke must be constructed at the level recommended by the turbine manufacturer.

Technical specifications

For Pelton or Michell-Banki turbines, the technical specifications of the tailrace channel are the same as for the conveyance channel.

For reaction turbines (Kaplan, Francis or pumps as turbines), the technical specifications of the tailrace channel also depend on the hydraulic characteristics of the draft tube, the height of the dyke, the slope before and after the dyke.

CHAPTER 14
Commissioning

This chapter deals with to the commissioning of the civil works and electromechanical equipment; it does not include transmission and distribution lines, which do not fall within the scope of this book.

Hydraulic tests on civil works

Intake

Each of these components needs to be tested twice: once in the dry season and once in the rainy season.

Guiding walls

Check stability and resilience to bending, sliding and settling, making sure they remain upright, with no cracks, dents or shifting from their original position.

Intake aperture

Check:
- The dimensions are adequate for the design flow during the dry season, acting as a spillway.
- In the rainy season, the intake aperture takes the design flow and the excess flow.
- The safety screen prevents stones or floating debris (stalks, branches, etc.) from passing through the aperture.
- The ledge of the intake aperture is adequate, as it prevents debris carried by the river from entering the headrace.

Fixed weir

Check:
- The edges of the columns are not damaged by the debris carried by the river during the rainy season.
- The columns are resilient to bending and to the pressure of the water during rainy seasons.
- The stop log grooves efficiently support the planks that form the removable part of the weir.

- The height of the trash rack does not cause the river to overflow upstream.
- No water spills over onto the guiding walls.

Removable part of the weir

Check:
- The planks are sufficiently resilient to dampness, to the pressure of the water and to the debris carried by the river during the rainy season.
- During the dry season, the water is held back by the weir and conveyed to the intake aperture and the excess flow runs over it with no difficulty.
- During the rainy season, it acts as a thin-edged spillway.
- The stop logs are easy to remove and they regulate the intake flow during the rainy season.
- Sediments are cleared easily when the stop logs are removed and the water current flows.

Stonework

Check:
- It has withstood the impact of the water pressure produced by the weir (fixed and removable) during the rainy season.
- The debris carried by the river during the rainy season has not damaged the stonework surface.
- There are no cracks on the surface.

Headrace channel

Check:
- The design flow plus the excess flow do not spill over the free border and that the spillway and the overflow channel divert the excess flow during the rainy season.
- No sediments are formed in this section.
- The walls and floor show no cracks or settling and the slope is even.
- The construction joints show no deterioration, nor have they shifted.
- The lining and waterproofing remain unchanged.

Side spillway

Check:
- It evacuates the excess flow satisfactorily.
- The overflow channel conveys the excess flow without eroding the adjacent land.

Coarse settling basin

Check:
- It evacuates the anticipated flow containing sand and thick particles through the flood gate.
- The desilting channel does not erode the surrounding land.

Conveyance channel and spillways

Check:
- The flow is the same as the design flow, applying the floater method or the salt solution method.
- The walls show no cracks, settling or water filtration.
- The expansion joints remain in place. The side spillway evacuates the entire flow in cases of emergency if a temporary wooden dyke is placed in the channel.
- The overflow channel of the side spillway works normally without eroding the adjacent land.
- No sediments were formed along the course of the channel.
- The slope is consistent with the design.
- The culverts and ditches work normally.
- The raised part of the channel platform is of the right size, fulfilling its purpose.
- The covered areas of the channel protect it from mud-slips on the adjacent slope.

Silt settling basin

Check:
- The anticipated sedimentation of fine and thick material in the intake aperture and its evacuation through the cleaning and draining system is satisfactory.
- The speed of the water slows down when it enters the settling basin, settling takes place at a low speed and the outflow increases to the channel speed.
- No turbulence is evident in the water flow.
- The side spillway and the overflow channel evacuate all the water normally when the flood gate is closed.
- The walls and floor show no cracks or settling.

Silt basin and forebay tank together

In addition to the checks for the silt basin, verify that:
- The settling speed is maintained when entering the forebay tank.
- The side spillways and the overflow channel evacuate the water satisfactorily when the turbine valves are shut.

- The overflow channel of the forebay tank takes away the excess flow when the turbine is operating.
- The screen of the forebay tank blocks floating debris such as leaves and branches.
- In addition to providing security, the surrounding fence also provides privacy for operation and maintenance tasks.
- The water level remains constant when the turbine and generator are operating with a full load.

Penstock

Check:
- No seepage or leaks are evident in any of the unions.
- The anchors are stable, and have neither tilted, shifted nor settled.
- The penstock is completely buried or protected from changes in temperature (PCV piping). The protecting soil is stable.
- The expansion joints work satisfactorily, the forebay tank and anchors show no cracks or splits (steel piping).
- The vent pipe is protected and working properly, automatically dissipating the energy from the water in the penstock when required to do so.

Powerhouse

In terms of the structure, architecture and safety, check:
- No cracks or subsidence are evident in the foundations, walls, floors or roof.
- There is no evidence of damp on the walls, floors or pavements. The drainage ditch works normally.
- The distribution of the electromechanical equipment and the access to each unit is adequate, easing operation and maintenance work.
- It is safe enough to store the electromechanical equipment.
- It is not exposed to potential hazards such as rocks sliding down unstable slopes or flooding during the rainy season.
- Lighting and ventilation are adequate. Indoor and outdoor lights work normally at night.
- The sanitary fixtures are appropriate and fulfil their purpose.

Discharge channel

Check:
- It operates normally, with the necessary section and slope to evacuate the water from the turbine in the case of an open channel and, in the case of a closed channel (PVC pipe), the diameter and the slope are adequate and equipped with valve boxes.

- The discharge head in the river is higher than the maximum level of the river during the rainy season.

Foundation for the electromechanical equipment

Check:
- The foundation bases satisfactorily withstand the pressure of the turbine and generator when they are operating. There is no evidence of cracking or settling.
- The level of the foundations for the equipment is adequate for the assembly and operation of the equipment.

Protocol for operating tests of the electromechanical equipment in the powerhouse

Turbine

Verify that all rotating elements were spinning normally:
- The shaft spins evenly, reaching the design RPM, in accordance with the design head and flow.
- The pulley spins in one plane and is dynamically balanced.
- The belts are adequately taut and do not heat, skid or slip.
- The seal works properly; no leaks through the shaft are evident.
- The runner spins evenly without brushing against the nozzles.
- The roller bearings do not heat and the shaft produces no sounds or screeches, showing the selection and installation are satisfactory.
- When all the nozzles are open the turbine does not speed up, the controller and the ballast load work well.
- Water flows through the ballast load.

For the non-rotating elements, check:
- The valves work normally when open or shut. The water flows normally and there is no evidence of leakage through the packing washers.
- The nozzles work normally, there are no stones or obstacles preventing the outflow, in accordance with the design. The volume of water is consistent with the design.
- There are no leaks in the unions of the inlet pipe or in the penstock coupling accessories.
- The whole turbine is sturdy and works well, with no vibration and the screws remain tight.

Generator

Check:
- The pulley is well adjusted, with no sagging.

- There is no vibration in the generator. The turbine and generator pulleys are well balanced, levelled and aligned, the transmission system is appropriate and the internal structure of the generator is of a reliable quality.
- The bearings do not heat. The belts are adequately taut and the bearings are well-greased.
- The bearings make no sounds, showing they have been selected and adjusted properly.
- The voltage produced by the generator is consistent with the design. The cable connections and AVR calibration are correct, reaching the RPM indicated by the manufacturer.

Control panel

Check:
- When the generator is fully operating, the control panel correctly marks the frequency, the voltage and the current intensity in all three phases.

Electronic load controller

Check:
- When the turbine valves are opened progressively, the turbine runs at a standard speed.
- The ballast load dissipates energy and the voltage remains constant.
- The water flowing in and out of the resistance tank is constant. Neither the resistances nor the cooling water heat up.
- The bypass valve and the diversion pipe from the turbine inlet pipe to the resistance tank permit a constant flow of water, with no obstructions.

Acceptance certificate

Activities prior to the delivery of the works

- The implementer informs the owner that, before the works are formally handed over, it is important to go over the entire infrastructure of the hydroelectric system in the presence of an expert, to make a note of any observations that may arise.
- Through the hired expert and within a reasonable time, the owner informs the works implementer of the observations made.
- Within a reasonable time, the works implementer studies the observations and informs the owner that they have been dealt with. At the same time, he sends a written invitation establishing the time and date of the formal delivery of the works.

Activities required during the delivery of the works

- On the specified day, the operation of the hydroelectric system is tested by the owner, together with one or more experts and the implementer, making sure that the protocol tests are satisfactory, in keeping with the technical design of the entire installed infrastructure, or only of the part constructed. These tests are recorded on pre-designed forms which form part of the acceptance certificate.
- Once all the protocol tests have been successfully completed, the MHS must be left operating for at least 24 hours, under the direction and supervision of a duly trained operator.
- The next day, both parties sign the acceptance certificate, whereby the owner accepts the entire infrastructure in good working order and takes over full responsibility for the satisfactory operation and maintenance of the system.

Guidelines for the acceptance certificate

The following information is usually included:
- date;
- time;
- place;
- name of the hydroelectric plant;
- names and positions of the members of the acceptance committee;
- verification and results of the tests conducted on each component;
- verification of the normal operation of the entire system for at least 24 hours;
- the owner's commitment to employ trained staff for the proper operation and maintenance of the system;
- confirmation that the implementer is delivering the works in good condition, endorsed by experts, and that the owner accepts the works to his or her entire satisfaction;
- the signatures of both parties and of other local authorities invited to attend the acceptance ceremony.

Figure 14.1 Example of an acceptance certificate

At 10 o'clock on the morning of 5 October 2006, a meeting was held in the meeting room of the District Municipality of Chetilla, province and department of Cajamarca, Peru, attended by the Mayor, Mr. ...... on behalf of the municipality; power generating and power distribution experts Messrs. .... and...... on behalf of the Chetilla MHS works implementer; and construction foreman Mr. .... Then they all went to the site of the micro hydroelectric plant to verify its normal operation and carry out the formal handover and acceptance of the micro hydro scheme.

After visiting the facilities and carrying out the corresponding tests, the abovementioned people certified that:

1. The civil works – comprising a weir with removable stop logs, a tailrace channel, a conveyance channel, a settling basin, a forebay tank, complementary works, a penstock, a powerhouse and a discharge channel – were all in good working order.

This infrastructure had been built in accordance with the technical specifications, with a few modifications of a structural nature. The corresponding hydraulic tests were conducted on each of the components, with satisfactory results.

2. The electromechanical equipment – comprising a .... type turbine manufactured by ...., a 220/380 V, 60 Hz three-phase generator manufactured by ...., a control panel, an electronic load controller and the ballast load, with measuring instruments, alarms and protected by the manufacturers .... – is working normally, generating the designed electrical power. The units, manufactured locally and abroad, are all of a good quality and international standard. It was verified that they were purchased and installed in accordance with the project's technical specifications and characteristics and optimum results were obtained from the protocol tests carried out on each unit.

3. The electrical distribution system has passed all the protocol tests; the materials, equipment and installation respond to the technical standards and characteristics of low and medium voltage systems.

4. It was verified that, after operating for 24 hours, power was being generated and distributed normally, with no interruptions, voltage dips, etc.

5. The district municipality has guaranteed that it will operate the plant properly, with staff duly qualified in the operation and maintenance of the system.

6. The MHS operators have proved that they are fully trained to manage the system properly; they have the necessary tools, measuring instruments, security equipment, an operation and maintenance handbook. They make up an effective management team to provide a good service.

Consequently, by way of delivering and accepting the works in optimum condition, both parties and the local authorities present hereby place their signatures on this document.

....
For the Implementing Company

....
Expert 1

....
Expert 2

....
Mayor

Signatures of other authorities

CHAPTER 15
Staff training

Operation and maintenance staff of a micro hydroelectric power plant are selected based on certain requirements such as their knowledge, behaviour, age, creativity, local residence and their participation in the construction of the MHS.

They receive theoretical and practical training on specific topics, selected in accordance with the works implemented and put into operation.

They are trained by a team of technicians widely experienced in the construction of civil works and the manufacture and assembly of electromechanical equipment, as well as in the design and installation of medium and low voltage distribution systems.

The training sessions are divided into the following topics.

Basic knowledge

1. Components of the MHS: civil works, electromechanical equipment, distribution systems and general layout.
2. Civil works in MHSs.
   - Components of the civil works and the function of each one: intake weir, headrace channel and conveyance channel, spillway and overflow channel, settling basin(s), forebay tank, tailrace channel.
   - Layout of each element of the civil works.
3. Electromechanical equipment
   - Electromechanical equipment components: turbine, generator, electronic load controller, resistance tank and control panel.
   - Main parts of the turbine. Layout of the turbine (in pieces).
   - Functions of the turbine and each of its elements and of the power transmission system.
   - Concept of the electric power elements: voltage, current, resistance and power. Measuring units.
   - Generator and its main parts. Layout of the generator.
   - Function of the generator.
   - Main parts of the control panel.
   - Function of the control panel.
   - Reading of the measurement instruments: voltmeter, ammeter, multimeter.
   - Main parts of the electronic load controller.
   - Function of the electronic load controller.
   - Main parts of the ballast load.
   - Function of the ballast load.

4. Cables between the generator and the control panel, electronic load controller and ballast load tank.
5. Importance and parts of the earthing system. Layout.
6. Main parts and function of the primary grid with medium voltage and of the distribution with low voltage.
7. Safety standards and equipment.
8. Human relations within the community: organization structure, representative authorities, community institutions, channels of communication and information, and inter-personal relations.

Training in the operation of the hydroelectric system

1. In civil works:
   - Regulation of the intake flow during the low flow stage and the high flow stage, in the intake weir, in the headrace channel and in the conveyance channel.
   - Clearing sediments from the settling basin and forebay tank.
   - Regulation of the flow in the settling basin and forebay tank.
   - Filling the penstock with water to run the turbine.
2. In the electromechanical equipment:
   - Procedure for starting up the turbine, generator, electronic load controller, resistance tank and tailrace channel.
   - Procedure for interrupting the electric power generating and distribution process.

Basic preventive maintenance of the hydroelectric power system in:

- Civil works
- Electromechanical equipment
- Connections
- Distribution systems

Additional specialized training

- Specialized preventive maintenance of the different components of the hydroelectric system
- Equipment, measuring instruments, basic tools and materials required for the preventive maintenance of the components of the hydroelectric system
- Registration of the maintenance works and/or repairs, causes and solutions
- Daily records of the energy consumed in off-peak hours and peak hours

References and Bibliography

ACI (1998) *International Structural Engineering and Construction Congress*. Peruvian Chapter, ACI, Lima, Peru.

Agustoni, Cesare (1972) 'Hydraulic machinery for the Italian hydroelectric plants'; issued by the Organizing Committee of the 6th Symposium of the International Association of Hydraulic Research – Section Hydraulic Machinery, Equipment and Cavitation, Rome.

Brown, A.P., R.C. Edwards, J. Gilliam and W.J. Langley (1995) 'Discharge monitoring using salt gulp and automated stage measurement techniques'; Dulas Engineering Ltd., Proceedings of the Conference 'Hydropower Into The Next Century, Potential–Challenges–Opportunities', Barcelona, Spain.

Goldemberg, J. (2000) *World Energy Assessment 2000: Energy and the Challenge of Sustainability*, UNDP / UN-DESA / World Energy Council, New York.

Harvey, Adam (1993) *Micro Hydro Design Manual, A guide to small-scale water power schemes*. Practical Action Publications, UK.

Husebye, S. (1995) 'Hydropower and the environment – the work of IEA'. *Hydropower & Dams*, May 1995, Aqua-Media International Ltd., Sutton, UK.

Indacochea, R. De S. Enrique (1979) 'Problemática del Desarrollo de la Tecnología de Microcentrales Hidroeléctricas y su contribución a la Electrificación Rural', (Development Problems of the Micro Hydroelectric Scheme Technology and its contribution to Rural Electrification). Institute for Industrial Technological Research and Technical Standards) (ITINTEC), Lima, Peru.

ITDG (1988) 'Agua, energía y desarrollo rural', seminario taller 'Hidroenergía y desarrollo rural' (Water, energy and rural development, Seminar Workshop on Hydro power and rural development), Cusco, Peru.

ITDG Peru (1992) *Manual para la construccion de canales por el metodo de las perchas (Handbook on the construction of canals with the wooden frame method)*. ITDG Peru, Lima, Peru.

ITDG Peru (1996) *Manual de mini y microcentrales hidroeléctricas- Una guía para el desarrollo de proyectos (Handbook on mini and micro hydroelectric plants: A project development guide)*. ITDG Peru, Lima, Peru.

International Energy Agency (IEA) (2006) *World Energy Outlook 2006*. IEA, Paris.

JUNAC (1984) *Farmer Manual Silvo* Volume 11, Rural Equipment. Cajamarca, Peru.

Lauterjung, Helmut and Gangolf Schmidt (1989) *Planning of Water intake structures for irrigation or hydropower*. Vieweg, Wiesbaden, Germany.

Marchegiani, Ariel R. (2006) *Renewable Energy in Developing Countries*. Engineering faculty, Nacional University of Comahue, Buenos Aires, Argentina.

Nozaki, Tsuguo (1981) JICA Expert, *Guía para la Elaboración de Proyectos de Pequeñas Centrales Hidroeléctricas destinadas a la Electrificación Rural de Perú (Anexo Suplementario) (Guide for Small Hydroelectric Projects aimed at Rural Electrification in Peru (Supplementary Annexe)*. March 1981, Lima, Peru.

Nechleba, Miroslav (1957) *Hydraulic Turbines, their Design and Equipment*. ARTIA, Prague, Czechoslovakia.

OLADE (1981) *El Potencial Hidroeléctrico Alterntiva Energética y Desafío Industrial y Financiamiento para América Latinañ* (Hydroelectric Potential, Energy Alternatives and Industrial and Financing Challenge for Latin America – series) OLADE No 18, first edition.

Rodríguez Sánchez, Luis (1998) 'Obras civiles en microcentrales hidroeléctricas' (Civil works in micro hydroelectric plants). *HIDORED*, January 1998, Lima, Peru.

Sánchez Campos, Teodoro (1988) 'Behaviour of Centrifugal Pumps Running as Turbines', MSc thesis, University of Reading, Reading, UK.

Sánchez Campos, Teodoro, and Luis Rodríguez Sánchez (2001) ITDG Peru, 'Fondo de promoción de microcentrales hidroeléctricas, un modelo financiero con subsidios y asistencia técnica, Artículo presentado al IX ELPAH (Promotion fund for micro hydroelectric plants, a financial model with subsidies and technical assistance', article submitted to *IX ELPAH*, Neuqén, Argentina.

Sánchez, Teodoro (2006) 'The critical factors for successful stand alone energy schemes in Peru', unpublished PhD thesis, Nottingham Trent University, Nottingham.

Terán, Rubén (1998) 'Diseño y construcción de defensas ribereñas (Design and construction of river wall protection)', Escuela Superior de Administración de aguas (School for Water Management), Arequipa, Peru.

UNDP (United Nations Development Program) Web page http://www.undp.org/energy/

UNIDO (2004) *Hydrodynamic Design Guide for Small Francis and Propeller Turbines*. Vienna, Austria.

World Bank (1996) *Rural Energy and Development, Improving Energy Supplies for Two Billion People*. Washington D.C.

World Bank (2003) 'Village Energy Solutions for Remote Areas of Brazil', ESMAP Concept Note, http://www.esmap.org/filez/activity/312007114922_LACBrazilVillageEnergySolutionsRuralAreas.pdf

EXAMPLE 1
Calculation of a mixed intake (see Chapter 2)

This example describes the particular case of Las Juntas MHS (Jaen, Peru, 1998), with the following river-related field data.
- Width: 7.80 m (dry season)
- Average head during the dry season 0.30 m
- Slope of the river along the section of the intake: 1.54%
- Slope of the right bank (downstream): semi-split rock, angle of the slope with the horizontal 37°
- Slope of the left bank (downstream): clayey soil with a little sand and pebbles, with a 75° slope up to 1.42 m high, then rice fields.
- Head during maximum high flow, 0.65–0.70 m
- Measurement in dry season: 1.150 m³/s
- Diagram of the cross-section (Figure A1.1)

With the field data, the following was determined for the design of the guide walls.
- Material: coarse concrete, this being a monolithic structure resilient to abrasion and erosion, durable for at least 30 years. Taking advantage of good quality local aggregates (stone, sand and reinforced concrete) reduces the cost of the material and transport.
- The foundation soil on the right bank is rocky, with an admissible bearing capacity of 4 kg/cm². The soil on the left bank is clay with a little compacted sand and pebbles of varying diameters, with a minimum admissible capacity of 2 kg/cm².
- The intake aperture is situated in the wall on the right bank, which is an appropriate location for the construction of the conveyance channel and other components.

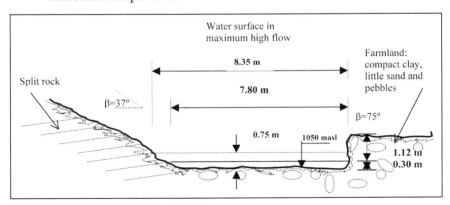

Figure A1.1 Cross-section of the river for the intake design

- The design for the wall was based on the left bank, taking into account the more unfavourable stability conditions.
- Recommendations of the hydrological study: most representative volume during rainy seasons: 13 to 14 m³/s, silt, scouring, etc.
- Average speed of the water during the rainy season: 2.9 m/s

Design of the wall

Selection and sizing of the cross-section

Calculation of H_1

H_1 = thickness of the cover slab + H_{weir} + $H_{max\ high\ flow}$ + freeboard

Thickness of the cover slab = 0.20 m (0.15 to 0.20 m was assumed, in accordance with the volume of the river).

H_{WEIR} = window ledge + river flow in normal conditions:
0.20 + 0.30 = 0.50 m, 0.60 m can be considered (the window ledge is between 0.12 and 0.25 m so that no large stones cover the intake aperture).

$H_{max\ high\ flow}$ is deduced from the spillway formula (h):

$$Q_{max} = \frac{2}{3} \times \mu \times b \times \sqrt{2g}\left[\left(h+\frac{V^2}{2g}\right)^{3/2} - \left(\frac{V^2}{2g}\right)^{3.2}\right],\ \text{using field data}$$

$$13\ m^3/s = \frac{2}{3} \times 0.6 \times 7.80 \times \sqrt{2 \times 9.8}\left[\left(h+\frac{2.9^2}{19.6}\right)^{3/2} - \left(\frac{2.9^2}{19.6}\right)^{3/2}\right]$$

Clearing $h = 0.70\ m = H_{max.high\ flow}$

Freeboard = 0.30 (assumed)

Then: $H_1 = 0.20 + 0.60 + 0.70 + 0.30 = 1.80\ m$; Figure A1.2 shows the first attempt:

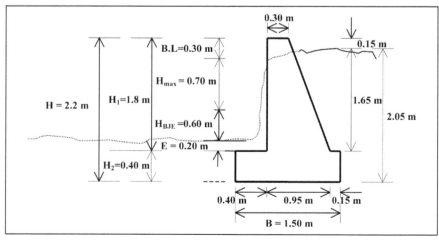

Figure A1.2 Preliminary sizing of the wall

CALCULATION OF A MIXED INTAKE

Intervening forces

Intervening forces analysed by weight: wall, soil and thrust of the earth on the wall.

Sub-pressure forces at the base of the wall and of the water contained in the soil are not analysed. They should be eliminated during the construction process, through a drain.

From Figure A1.3, the weight of the wall is determined: $Pm = P_1 + P_2 + P_3$
Specific weight of the coarse concrete = 2.2 T/m³ (known fact)
$P_1 = 1.50 \times 0.40 \times 2.2 = 1.320$ T. Acting at 0.75 m from A
$P_2 = 0.30 \times 1.80 \times 2.2 = 1.188$ T. Acting at 0.55 m from A
$P_3 = [(0.65 \times 1.80)/2] \times 2.2 = 1.287$ T. Acting at 0.916 m from A

Weight of the soil on the wall: $Pm = T_1 + T_2$
Obtained from Table 2.1: Specific weight of the soil (dry clay) = 1.6 T/m³
$T_1 = [(0.596 \times 1.65)/2] \, 1.6 = 0.786$ T. Acting at 1.15 m from A
$T_2 = 0.15 \times 1.65 \times 1.6 = 0.396$ T. Acting at 1.425 m from A
Active thrust of the soil:

$$E_t = \frac{1}{2} \times K_a \times \gamma \times H^2$$

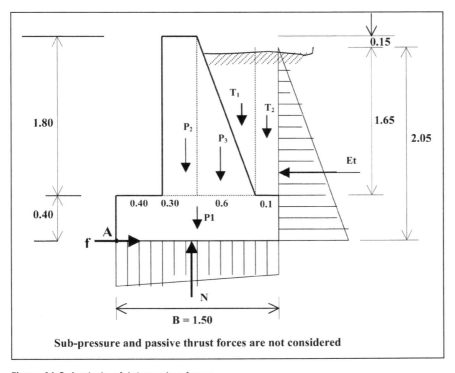

Figure A1.3 Analysis of intervening forces

Calculation of coefficient K_a

$$K_a = \cos\theta \times \frac{\cos\theta - \sqrt{\cos^2\theta - \cos^2\phi}}{\cos\theta + \sqrt{\cos^2\theta - \cos^2\phi}}$$

According to field data and the values from Table 2.1.
$\theta° = 0$ (the ground in contact with the wall does not form an angle with the horizontal)
$\phi = 45°$ (internal friction angle of the clay)
Replacing data:

$$K_a = \frac{1 - \sin\phi}{1 + \sin\phi}$$

$K_a = 0.172$

Then: $E_t = \frac{1}{2} \times 0.172 \times 1.6 \times 2.05^2$

$E_t = 0.578$ T acting at 0.68 m from A

Soil reaction:
$N = P_1 + P_2 + P_3 + T_1 + T_2 = 4.977$ tons

Bending stability:

Condition to be fulfilled: $\frac{\Sigma StabilizingM}{BendingM} \geq 2$

Stabilizing and bending moments are taken with respect to point A
$\Sigma Stabilizing\, M = 1.32 \times 0.75 + 1.188 \times 0.55 + 1.287 \times 0.916 + 0.786 \times 1.15 + 0.396 \times 1.425 = 4.289$ MT
Bending $M = 0.578 \times 0.683 = 0.395$ MT

Verifying: $\frac{4.289}{0.395} \geq 2$

$10.85 \geq 2$ ... OK

Sliding stability:

Condition: $\frac{f}{E_t} \geq 2$

Force of friction $f = \mu N$ (in Table 2.2, $\mu = 0.45$)

Verifying: $\frac{0.45 \times 4.977}{0.578} \geq 2$

$3.87 \geq 2$ ... OK

Stability of the foundation soil:
Condition to be fulfilled: $\sigma_{bearing\ capacity} > \sigma_{max} > \sigma_{min}$

From Table 2.2: $\sigma_{bearing\ capacity} = 1$ kg/cm²

$$\sigma_{min} = \frac{N}{B}\left[1 - \frac{6 \times e}{B}\right]$$

$$\sigma_{max} = \frac{N}{B}\left[1 + \frac{6 \times e}{B}\right]$$

$$e = \frac{\Sigma M_o}{N}$$

ΣM_o = Sum of the moments of all vertical and horizontal forces with respect to the centre of base B of the wall footing

$$e = \frac{1.32 \times 0 - 1.188 \times 0.20 + 1.287 \times 0.166 + 0.786 \times 0.401}{4.977} +$$

$$\frac{0.396 \times 0.675 - 0.578 \times 0.683 + 2.239 \times 0}{4.977}$$

$e = 0.0328$ m
Replacing values:

$$\sigma_{min} = \frac{4.977}{1.50}\left[1 - \frac{6 \times 0.0328}{1.50}\right] = 2.88 \text{ T/m}^2 = 0.29 \text{ kg/cm}^2$$

$$\sigma_{max} = \frac{4.977}{1.50}\left[1 + \frac{6 \times 0.0328}{1.50}\right] = 3.75 \text{ T/m}^2 = 0.38 \text{ kg/cm}^2$$

Verifying : 1 kg/cm^2 > 0.38 kg/cm^2 > 0.29 kg/cm^2 ... OK.

The dimensions assigned to the wall sufficiently comply with all the stability conditions to allow for bending stability. However, an attempt can be made to reduce the dimensions in order to reduce costs.

The new sizing (Figure A1.4) complies with all balance conditions and the cost is 28 per cent lower.

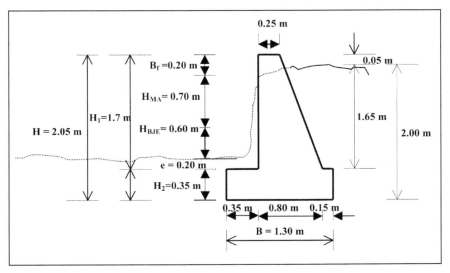

Figure A1.4 Final resizing of the wall

With the same procedure and bearing in mind the characteristics of the concrete and soil, the following results were obtained.

Weight of the wall:
P1 = 1.001 T, acting at 0.65 m from end A and 0.000 m from the centre of the base.
P2 = 0.935 T, acting at 0.475 m from end A and 0.175 m from the centre of the base.
P3 = 1.028 T, acting at 0.780 m from end A and 0.13 m from the centre of the base.
Weight of the soil:
T1 = 0.704 T, acting at 0.97 m from end A and 0.32 m from the centre of the base.
T2 = 0.396 T, acting at 1.225 m from end A and .575 m from the centre of the base.
Reaction of the soil: N = 4.064 T
Force of friction: f = 1.828 T
Active thrust of the soil: E_t = 0.550 T, acting at 0.666 m from A and from the centre of the base.
Stabilizing moment: ΣM_e = 3.068 MT
Moment with respect to the centre of the base: ΣM_o = 0.061 MT
Eccentricity: e = 0.015 m
Bending verification: 8.45 ≥ 2
Sliding verification: 3.32 ≥ 2
Foundation soil verification: 1 kg/cm² > 0.33 kg/cm² > 0.29 kg/cm²

Design of the intake aperture (Figure A1.5)

This consists of calculating the dimensions of the intake aperture through which the project design flow will enter. Orifice formulae are applied here, bearing in mind the river head during both the dry season and the rainy season and, at the same time, determining the dimensions of the headrace channel for both cases, so that the water intake is safely conveyed to the first coarse settling basin or silt basin.

Calculating the window ledge d:

- The height of the intake aperture with respect to the level of the (improved) river bed.
- Based directly on the size of the silt carried by the river or the sediments it leaves when the rainy season stops. This usually varies from 15 to 25 cm.
- Make sure that most of the silt carried by the river passes underneath the window ledge, allowing the water to enter freely on the side with no obstructions.
- In this case, a 0.20 m ledge was chosen, based on the observation that the size of the stones and sediments left on the river shore reached 15 cm.

CALCULATION OF A MIXED INTAKE 257

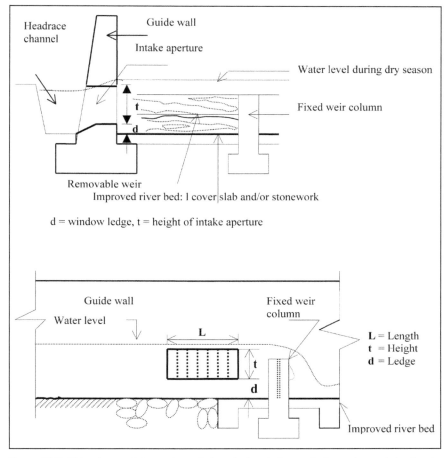

Figure A1.5 Side view and front view of the window ledge and dimensions of the intake aperture

Calculating L and t

These measurements were calculated bearing in mind the field data and the hydraulic theory of the spillway and the submerged orifice.
Field data on the river during the dry season:
$Q_r = 1.150$ m³/s (flow)
$B = 7.80$ m (width)
$Y_r = 0.30$ m (normal head)
$Y'_r = 0.70$ m (head of the river during the high-dry season)
Design data:
$Q = 0.625$ m³/s (catchment flow: 0.600 for the turbine, 0.020 for the permanent outlet in the settling basin and 0.005 for other uses)
The solution is as follows:

1. Determining the height of the weir

In rivers or watercourses in highland and jungle fringe areas, the head during low flow and high flow usually varies between 0.20 and 0.70 m on average, and the volume varies from 0.050 to 12 m³/s. Therefore, the height of the weir bears a relation to the bottom ledge of the intake aperture:

$3d \leq H_{WEIR} \leq 3.5d$

It was previously determined that: d = 0.20 m. According to that ratio, it was established that $H_B = 3 \times 0.20 = 0.60$ m. In such conditions, the river water level upstream from the weir is 0.60 + river head = 0.60 + 0.30 = 0.90 m (see Figure A1.6).

2. Dimensions L and t, calculated for normal conditions (dry season) using the following formula:

$Q = C_d \times A \times V$, corresponds to the flow, which in this case goes through a submerged orifice,

where:
C_d = discharge coefficient of the submerged spillway. In this case, because the wall is wide, it was corrected by another factor. On average, a value of 0.6 was taken, bearing in mind that the edges of the intake aperture are not angular but rounded and flat, as shown in Figure A1.6.
$A = L \times t$
V = speed of the water intake through the orifice

Then: $Q = C_d \times L \times t \times \sqrt{2g(H_r - Y_c)}$ (1)

g = gravity acceleration 9.8 m/s²
H_r = head of the river water above the window ledge and upstream from the weir
Y_c = head of the headrace channel

$V = \sqrt{2g(H_r - Y_c)}$ = speed of the water entering the intake aperture

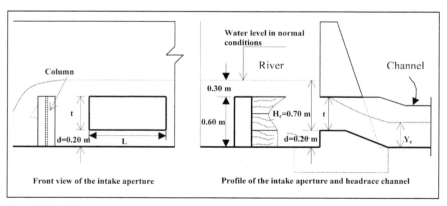

Figure A1.6 Dimensions L and t in the intake aperture

To replace data in formula (1),
It was assumed that $Y_c = t = 0.40$ m (continuity)
$Q = 0.625$ m³/s
$C_d = 0.6$
$H_r = 0.90 - 0.20 = 0.70$ m
$L = 1.07$ m, rounded to 1.10 m

$$0.625 = 0.6 \times L \times 0.40 \times \sqrt{2 \times 9.8 \times (0.70 - 0.40)}$$

With the assumed and calculated values, it was verified that the speed of the water intake will not erode the edges of the intake aperture.

Condition to be fulfilled: $V < 4$ m/s.

$V = \sqrt{2 \times 9.8 \times (0.70 - 0.40)} = 2.42$ m/s.

Therefore: $2.42 < 4$ ... OK.

3. Calculating width b and slope S of the headrace channel

It is recommended that the speed of the water in the channel should be $1.2 < V < 2.5$ so that no sediments remain, thus avoiding the erosion of its base and slopes.

For this case, a speed of 1.75 m/s was chosen, close to the intake speed.

With the continuity formula:
$Q = A \times V = b \times Y_c \times V$

b was found

$$b = \frac{Q}{Y_c \times V} = \frac{0.625}{0.40 \times 1.75} = 0.89 \approx 0.90$$

Calculation of the slope of the headrace channel:

With the Manning formula, $V = \frac{1}{n} \times R^{2/3} \times S^{1/2}$

Clear slope S; $S = \left(\frac{n \times V}{R^{2/3}}\right)^2$

V = Speed of the flow in m/s

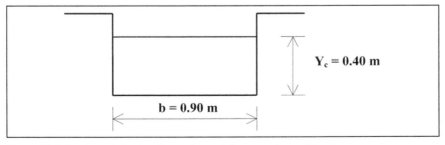

Figure A1.7 Section of the headrace channel (in normal conditions)

260 DESIGNING MINI AND MICRO HYDROPOWER SCHEMES

n = Manning coefficient for the concrete, polished finish = 0.016

R = hydraulic radius; $R = \dfrac{\text{Wet area}}{\text{Wet perimeter}}$

Wet area = $b \times Y_c$ = 0.90 × 0.40 = 0.360 m²
Wet perimeter = $b + 2Y_c$ = 0.90 + 0.80 = 1.70 m
Then, R = 2.1176
Therefore:

$$S = \left(\dfrac{0.016 \times 1.75}{0.2117^{2/3}}\right)^2 = 0.006 = 6 \text{ m/km}$$

Thus, the dimensions of the weir, the intake aperture and the headrace channel were determined, plus its slope for normal conditions (see Table A1.1).

Table A1.1 Results obtained

Dimensions/Characteristics	Weir	Intake aperture	Headrace channel
Bottom ledge	—	0.20	—
Height	0.60	0.40	—
Width	—	1.10	0.90
Head	—	—	0.40
Slope	—	—	0.006

During high flows in the rainy season, the flow in the headrace channel will increase due to the increase in H'_r. This channel must be prepared for this variation, calculating the new head Y_c', maintaining the value of b.

4. Calculating the head of the headrace channel for high flow conditions

In this case, $H'_r = H_B$ + high flow river head − d = 0.60 + 0.70 − 0.20 = 1.10 m
These measurements are represented in Figure A1.8.

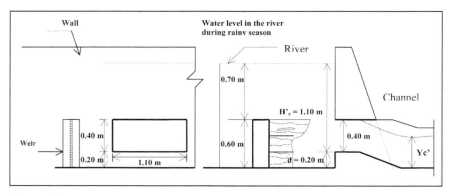

Figure A1.8 Water level and hydraulic characteristics in the intake aperture and the headrace channel during the rainy season.

a) $Q = L \times t \times C_d \times \sqrt{2g(H_r' - H_c')}$

With this formula, the volume flowing from the river to the channel through the orifice can be calculated.

H_r' = head or water level above the window ledge during the rainy season (upstream from the weir)

Y_c' = head of the headrace channel during the rainy season

The other formula corresponds to the flow conducted by the headrace channel during the rainy season, which is the same continuity formula

$Q = A \times V$

b) $Q = A \times \dfrac{1}{n} \times R^{2/3} \times S^{1/2}$

In this formula, the values of A and R are expressed, based on Y_c':

$A = b \times Y_c' = 0.90\, Y_c'$

$n = 0.016$

$R = \dfrac{A}{p} = \dfrac{0.90 Y_c'}{0.90 + 2 Y_c'}$

$S = 0.006$

The following are obtained by replacing values in formulas (a) and (b):

(a) $Q = 1.10 \times 0.40 \times 0.6 \times \sqrt{2 \times 9.8 \times (1.10 - Y_c')}$

(b) $Q = 0.90 \times Y_c' \dfrac{1}{0.016} \times \left(\dfrac{0.90\, Y_c'}{0.90 + 2\, Y_c'} \right) \times 0.006^{1/2}$

These two equations are solved by giving values to Y_c' until the same value of Q is obtained in both equations.

The values of Y_c' are higher than Y_c:

Results obtained:

Table A1.2 Results obtained

N° of test	Y_c' (m)	a) Q m³/s	b) Q m³/s
1	0.48	0.920	0.793
2	0.50	0.905	0.837
3	0.52	0.890	0.881
4	0.523	0.887	0.887

As can be appreciated, the third value of $Y_c' = 0.52$ m, gives approximate values of Q, differing only by 0.009 m³/s

To obtain more accurate values, a fourth test was carried out with $Y'_c = 0.523$, thus obtaining the same value for Q.

For practical reasons, $Y_c' = 0.53$ m was considered, bearing in mind that this head has increased by 13 cm. The freeboard of the channel should be added to this to prevent it from overflowing (see Figure A1.9).

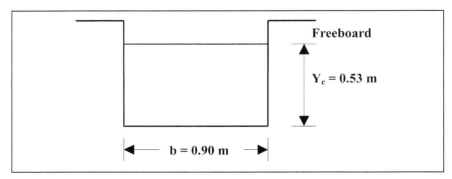

Figure A1.9 Section of the headrace channel (in high flow conditions)

Design of the fixed and removable weirs

The fixed weir comprises columns and a bracing slab, both of reinforced concrete, and a simple concrete cover slab.

The removable weir comprises wooden planks, to which a homemade preservative is applied (suet) to make it more damp-proof.

The column rests on the river bed through a footing and part of it is embedded in the bracing and cover slabs. The free part is short, to avoid the normal course of the water (see Figure A1.10).

Sizing of the weir-column with a stop log or gutter

For the outside or visible part

Section: Shaped like a polygon with the following characteristics:
- Inside angles 90° and 135°, to ease the normal course of the water
- The gutter for the stop log formed by b_1 and t_3 goes in the middle of the sides parallel to the course of the water.
- For symmetry and sturdiness, the following ratios are recommended :
  Sides that form right angles: $l = t_1 \times \sqrt{2}$
  Sides parallel to the gutter: $l' = 1.5 \times t_1$
  Width: $b = 2t_1$
  Gutter: $b_1 = t_3$ (in accordance with the breadth of the plank)
  Height of the weir: $2d \leq H_{WEIR} \leq 3.5d$
  The height should be as low as possible to ease the flow of the water, particularly during high flows.

For the concealed or embedded part

- Thickness of the simple concrete cover slab: $0.15 < e_1 < 0.20$
  This rests uniformly on the entire surface of the base or bracing slab.

CALCULATION OF A MIXED INTAKE 263

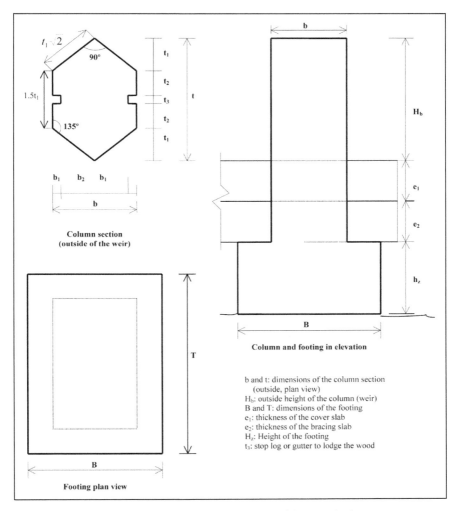

Figure A1.10 Geometric characteristics of the column with stop plank grooves

- Thickness of the reinforced concrete bracing slab: $0.12 < e_2 < 0.20$
  This rests uniformly on the entire surface of the river bed on the top part of the footings.
- Height of the footing: $0.30 < h_z < 0.45$
- It is necessary to know the bearing capacity of the soil $\sigma_t \geq 1$ kg/cm^2

With these recommendations, the dimensions of the column and footing for this case are shown in Figure A1.11.

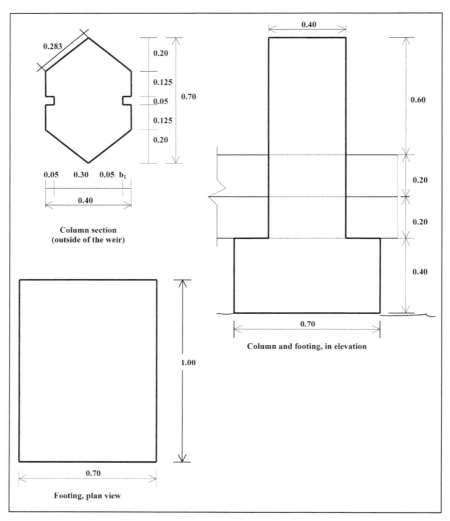

Figure A1.11 Dimensions of the column with stop log grooves

Steel in the column and footing (see Figure A1.12).

Sizing of the bracing slab (Figure A1.13)

L = total length of the slab (the same as the clear span between the guide walls)
$c = c_1 + c_2$ (width of the top base)
c_1 = length of the energy dissipation slab, passing the spillway.
$c_1 \geq 1.9\ Y_r'$ (Y_r' = head of the river during high flows)
$c_2 < c_1$
$h \geq e + c_3$

CALCULATION OF A MIXED INTAKE

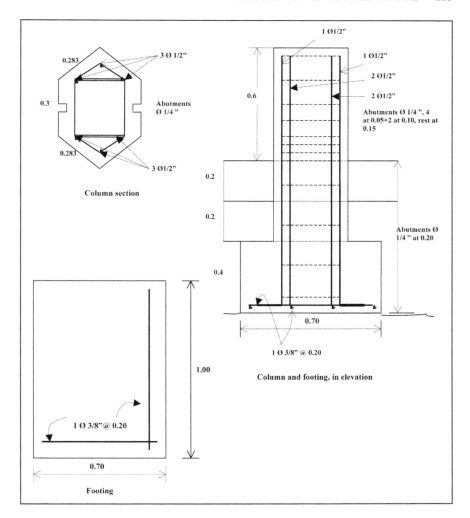

Figure A1.12 Steel distribution in column and footing

Angle α formed by the skirts of the slab may be 90° and 135°. The skirts help reduce sub-pressure forces.

The separation between the centre lines of the columns and between those and the adjacent walls must be the same, so that the pressure of the water is distributed evenly and directly over the weir. The planks must not be too long or too heavy. Between 1.50 and 1.85 m is recommended.

In accordance with the data on the river in the area of the intake (see Figure A1.1), the dimensions of the bracing slab and the columns are calculated as follows.

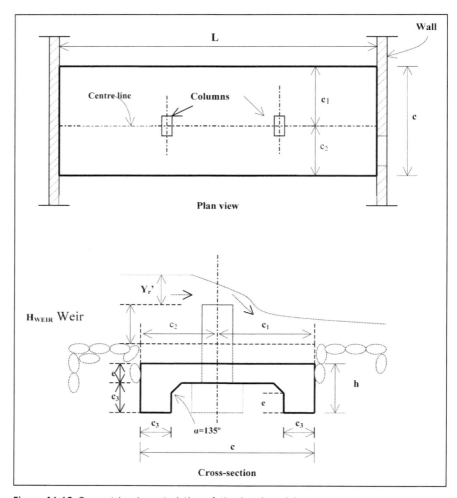

Figure A1.13 Geometric characteristics of the bracing slab

$L = 7.80$ m = clear span between the guide walls
$c_1 = 1.9 \times 0.70 = 1.33$. For security = 1.70 m
$c_2 < c_1$
$c_2 = 1.00$ m (assumed)
$c = c_1 + c_2 = 1.70 + 1.00 = 2.70$ m
$c_3 = 1.5\ e$ (e = thickness of slab = 0.20 m, assumed)
$c_3 = 1.5 \times 0.20 = 0.30$ m
$h \geq e + c_3$
$h \geq 0.20 + 0.30 = 0.50$ m

These dimensions and the longitudinal and crosswise steel reinforcements are shown in Figures A1.14 and A1.15.

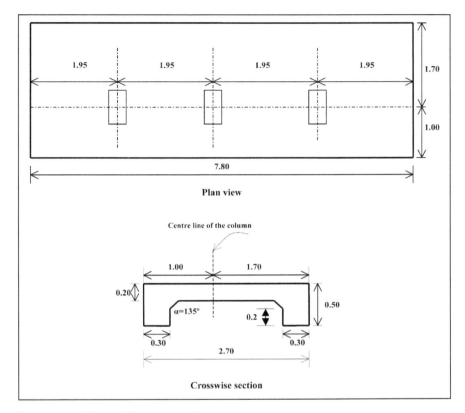

Figure A1.14 Sizing of the bracing slab

Cover slab

This is a prism with a rectangular base of the same dimensions as the base of the bracing slab. Its thickness may vary from 15 to 20 cm.

Its function is to protect the bracing slab, confine the columns, provide a perfectly uniform surface for the removable weir (planks), prevent the erosion of the river bed and help keep the guide walls from sliding and bending.

The material is simple concrete: cement and thick sand; split stones and medium-sized stones may be added, with an effective diameter no larger than 4" (10 cm) and up to 30 per cent of the total volume.

Stonework

This is a slab formed by stones arranged and stuck with cement and fine sand mortar, in the proportion indicated in the technical specifications.
- In the intake, the stonework goes on the river bed, before and after the cover slab, confined on the sides by the guide walls.

268 DESIGNING MINI AND MICRO HYDROPOWER SCHEMES

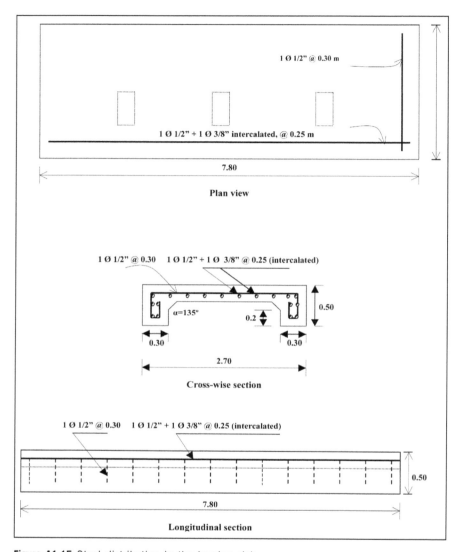

Figure A1.15 Steel distribution in the bracing slab

- It is important for the stonework slab to have skirts on both sides, so as to reduce sub-pressure forces considerably.
- Its main function is to prevent the river from eroding the river bed.
- Its thickness is usually ≥ 0.20 m
- Its length extends from the beginning to the end of the wall towards the cover slab.
- The width is the clear span between the guide walls.

- It is recommended that the skirt should be ≥ 0.50 m deep and its thickness and width should be at least 0.30 m.
- The stones should be washed, preferably rounded and fine grain. The river itself often provides them; porous or lightweight stones should be avoided.

Figure A1.16 shows the shape of its cross-section, which is very similar to that of the bracing slab.

Removable weir

The removable weir comprises planks that are wedged into the gutters of the columns and/or guide walls.

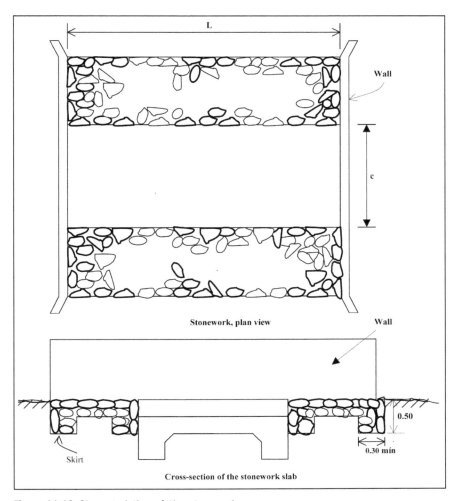

Figure A1.16 Characteristics of the stonework

During the dry season, the weir raises the level of the water above the ledge of the intake aperture; it evacuates the surplus flow as a submerged spillway during the dry season; it also works as a submerged spillway in the rainy season, but with a larger head.

It is removable so that, during the rainy season, the river head and the free passage of silt can be regulated by removing one or more planks; in dry seasons, all the planks can be removed for general cleaning and maintenance purposes.

The dimensions of the planks are determined when the width of the river and the number of columns are designed; however they must be verified bearing in mind the mechanical characteristics of the wood and the intervening forces.

Length p of the plank will be the same as the separation between the columns, including the length of the gutter at both ends, and approximately 1cm less so that it can be installed and removed easily.

Depth q of the plank is constant, and is approximately 0.5 cm less than width t_3 of the groove.

Width s of the plant is based on the height of the column. It depends on the number of planks to be fitted in each section.

In most cases, these planks are prepared in accordance with the dimensions established for the columns and the distance separating them. Commercially produced planks do not normally correspond the the measurements required.

It is recommended that the wood selected for the planks should have natural properties to prevent their structure from buckling.

Forestry experts recommend that wood with a basic density of more than 0.60 g/cm^3 should be used. The basic density is the relation between the weight of the dry wood and the volume when the wood is fresh.

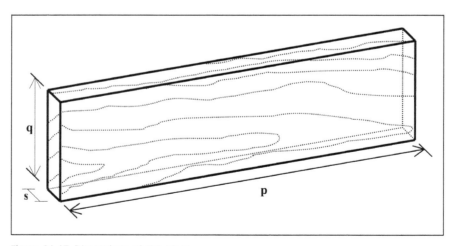

Figure A1.17 Dimensions of the plank

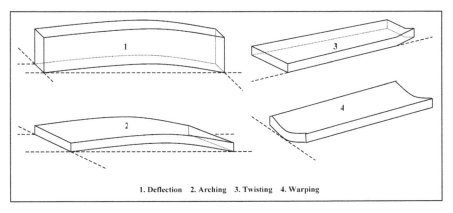

Figure A1.18 Wood distortion caused by drying

The wood used in intakes built in the highlands is eucalyptus; some studies indicate that its density is 0.58 g/cm³. In the Amazon areas there are several species of wood of high density. The results of the use of planks were good in terms of durability, functionality and low cost.

The durability factor depends on the force of the material carried by the river during the rainy season. If the material is angular, the top edge of the plank (crown) will deteriorate sooner than if it is struck by pebbles. In these and other conditions, because it is in contact with the water, eucalyptus will last approximately five years, but other more resistant species can last up to 10 years or even more.

In Peru the average cost of planks is about US$0.45 per square foot.

These weirs are usually only 0.50 to 0.70 m high. The thrust of the water is calculated with the following formula:

$E_a = \dfrac{1}{2} \times \gamma \times h^2$, varying between 125 kg/m and 245 kg/m for dry seasons,

when the water is free of silt.

In the rainy season, the specific weight of the water can be considered to be greater than 1 T/m³, as it will be carrying sand and/or silt, e.g. 1.2 T/m³. In this case, the force of the thrust increases from 150 kg/m to 294 kg/m.

To verify the preliminary sizing of the planks, use the following formulas, tables and subsequent steps, bearing in mind that the weir works as a beam embedded at both ends (stop log) (see Figure A1.19).

272 DESIGNING MINI AND MICRO HYDROPOWER SCHEMES

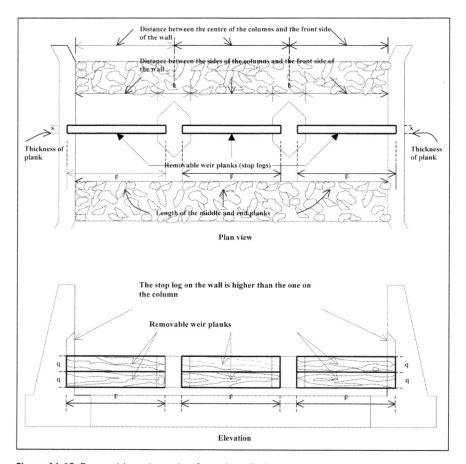

Figure A1.19 Removable weir made of wooden planks

a) $M_r = \dfrac{1}{6} \times \sigma_t \times q \times S^2$

M_r = resisting moment, in kg/cm
The resisting moment is equivalent to the sum of the acting moments.
Acting moments are usually produced by the distributed load and concentrated loads.
σ_t = stress due to buckling of the selected wood in kg/m^2 (see Table A1.3).
q = width of the plank in cm.
s = thickness of the plank in cm.

b) Bending moment and shearing force for a beam embedded at both ends (Figure A1.20):

CALCULATION OF A MIXED INTAKE

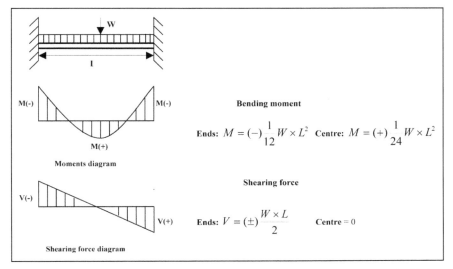

Figure A1.20 Diagram of the bending moment and shearing force for a beam embedded at both ends

Table A1.3 Mechanical characteristics of the timber frequently used in Peru

Timber	Young's modulus kg/cm^2	Expansion	Maximum permissible forces kg/cm^2		Bending	Shearing
			Compression			
			Parallel to the fibre	At right angles to the fibre		
Eucalyptus (dry)	100,000	105	110	28	150	12
'Tornillo'	90,000	75	80	15	100	8
Cypress	90,000	75	80	15	100	8
Common pine	90,000	75	80	15	100	8

Source: JUNAC, 1984

Procedure

The following data was used to calculate the weir for the Las Juntas MHS described above:
Gutter = 5 cm × 5 cm
Separation between the centre of each column and the front of the wall = 1.95 m
Width of column $b = 0.40$ m
Separation between the front of the wall and the side of each column (ends) = 1.95 − 0.20 = 1.75 m

- *Calculating length p of the plank*
  Length of end planks = 1.75 + 0.05 + 0.05 = 1.85 m
  Net length of plank = p = 1.85 − 0.015 = 1.835 m
  (Reduced by 1.5 cm for easier installation and removal purposes)
- *Breadth of the plank*
  s = width or depth of the gutter = 5 cm. Taking away 0.5 cm to ease its installation and removal, then s = 4.5 cm
- *Depth q of the plank*
  This is based on the number of planks to be placed on top of each other until the total height of the weir is obtained. Since the height of the weir varies between 0.50 and 0.70 m, the use of two or three planks is recommended. In this case, H_{WEIR} = 0.60 m, therefore two planks of q = 0.30 m (each) can be used.
- *Verification*
  Calculation of resulting moment M_r, acting on the plank:
  M_r = bending moment (distributed load) + concentrated forces moment
  The bending moment is defined by the permanent horizontal thrust of the water.
  It was considered that no concentrated horizontal forces act on the plank during dry seasons, but they do during the rainy season, when stones may strike the plank. These forces were taken into consideration to a certain extent by increasing the specific weight of the water from 1 to 1.2 T/m^3 in order to calculate the thrust of the water.
- Determining the thrust of the water E_a
- Calculating the bending moment:

$$M = \frac{1}{12} \times W \times L^2 = \frac{216 \times 1.75^2}{12} = 55.125 \text{ kg/m} = 5{,}512.5 \text{ kg/m}$$

L = 1.75 m (the clear span between the inner side of the column gutter and the wall)
Concentrated load moment = 0

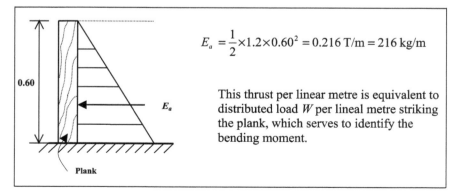

$$E_a = \frac{1}{2} \times 1.2 \times 0.60^2 = 0.216 \text{ T/m} = 216 \text{ kg/m}$$

This thrust per linear metre is equivalent to distributed load W per lineal metre striking the plank, which serves to identify the bending moment.

Figure A1.21 Bending moments over the slab due to the water thrust

- $M_r = M + 0 = 5{,}512.5$ kg/cm
- Verify that the pre-established dimensions of the plank respond to the bending and shearing forces, depending on the type of wood selected and the values of Table A1.3.

Verification of bending moment:

$$M_r = \frac{1}{6} \times \sigma_t \times q \times s^2 \quad \text{Clear } \sigma_t =$$

$$\sigma_t = \frac{6 \times M_r}{q \times s^2} = \frac{6 \times 5{,}512.5}{30 \times 4.5^2} = 54.44 \text{ kg/cm}^2 = \text{kg/cm}^2$$

A comparison with the values in Table A1.3 shows that the bending force of 55 kg/cm² is lower than the permissible bending force of eucalyptus (150), *Tornillo* (100), cypress (100) and common pine (100).
Therefore, the plank section is:
$q = 30$ cm, $s = 4.5$ cm, and $p = 1.835$ m for either of the four types of wood.

Furthermore, verifying the bending moment with length p of the planks in the middle section, their free length is 20 cm shorter; therefore, the bending moment will also be shorter. Consequently, the planks in the same section will be more resilient, which is justified as the speed of the water in the middle is faster than at the ends.

It is also worth stressing that, in an embedded beam, the negative moment was taken from the bending moment formulae, i.e. at the ends, as the absolute value is greater than the bending moment in the centre.

Verification of shearing force
Calculation of the shearing force:

$$V = \frac{W \times L}{2} = \frac{216 \times 1.75}{2} = 189 \text{ kg}$$

The verified section is 30 cm × 4.5 cm. Therefore, the shearing force is:

$$\sigma_c = \frac{V}{q \times s} = \frac{189}{30 \times 4.5} = 1.4 \text{ kg/cm}^2$$

Comparing this value with the values in Table A1.3: 1.4 kg/cm² is less than 8 and less than 12 kg/cm². Therefore, eucalyptus or any of the other three species may be used, i.e. *tornillo*, cypress or common pine. Of these four types of wood, eucalyptus and *tornillo* are the most durable.

Summarizing the calculation process, it can be observed that the breadth can be reduced, depending on the characteristics of the timber shown in Table A1.3. Furthermore, if it is desirable that the removable weir (planks) should break during extraordinarily high flows to allow the passage of silt and debris and prevent the water from overflowing over the guide walls, the breadth should be lower than the calculated figure.

Photograph A1.1 Mixed weir showing the guide walls, the fixed weir (columns) and the intake aperture

EXAMPLE 2
Calculation of a Tyrolean type intake

In the Manantial Eterno MHS (Tingo Maria, Peru), the topographic and hydraulic characteristics of the Santa Rosa river were obtained, as follows
Q_r = volume of the river = 0.450 m³/s (dry season)
Q_r' = average volume of the river in the rainy season 7.50 m³/s
Q = intake flow for generating electricity = 0.300 m³/s
h_o = 0.28 m (dry season)
h_o' = 0.65 m (rainy season)
Width of river: 9.90 m
Net width of the river for the weir and screen = 5.20 m
Slope of the river in the area of the intake = 12.95% (along 35 m)
Silt carried during the rainy season: thick sand and gravel (moderate)

Appropriate location for the settling basin: 15 m from the centre of the screen (left bank), where there is a favourable difference of 6.40 m in the level between the surface of the water in the river and in the settling basin. The soil on the left bank was stable, with typical jungle vegetation. The left bank had a hard rocky slope and, in the area of the settling basin, stable split rock and slopes with abundant natural vegetation (see Figure A2.1).

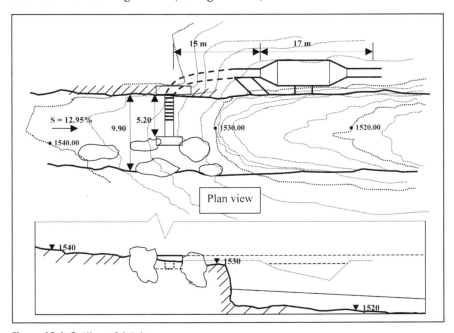

Figure A2.1 Outline of intake

1. The following formula provides the dimensions of the screen

$$Q = \frac{2}{3} \times c \times \mu \times b \times L \times \sqrt{2gh}$$

2. Calculation of each of the elements

$$c = 0.6 \times \frac{a}{d} \times \cos^{3/2} \beta$$

Select steel bars with a 1" (2.5 cm) diameter circular section, separated from each other by 2.5 cm.
Then:

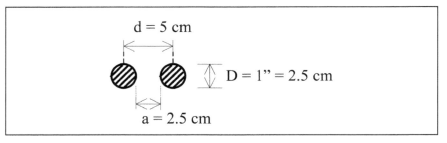

Figure A2.2 Distance between screen elements

Depending on the slope of the river in the section considered for the intake, the same slope applies to the screen. Therefore:
$\beta \approx \text{Arctan } 0.1295 = 7.378° \approx 7.5°$
With these values, calculate the value of c

$$c = 0.6 \times \frac{a}{d} \times \cos^{3/2} \beta = 0.6 \times \frac{2.5}{5} \times (0.9914)^{3/2} = 0.296 \approx 0.30$$

Calculation of h: $h = \frac{2}{3} \times k \times h_0$

In the table in Figure 2.13, $k \approx 0.93$, and according to field data, $h_o = 0.28$ m.
$h = 0.666 \times 0.93 \times 0.28 = 0.173$ m
According to the table showing contraction coefficients (see Figure 2.13), for the circular profile selected, the value of $\mu \approx 0.85$.

3. Replacing values in the Q formula, the following relation between b and L is obtained

$$Q = \frac{2}{3} \times c \times \mu \times b \times L \times \sqrt{2gh}$$

$$0.300 = \frac{2}{3} \times 0.30 \times 0.85 \times b \times L \times \sqrt{2 \times 9.8 \times 0.173}$$

$b \times L = 0.9593$
It was assumed that the value of b = half the width of the water surface in the river during the dry season: 5.20 m/2 = 2.60 m.

CALCULATION OF A TYROLEAN TYPE INTAKE

Then $L = 0.368$ m, and as part of the material gets trapped in the screen during the rainy season, 20 per cent more was considered. Therefore, the dimensions of the screen were:
$b = 2.60$ m and $L = 0.44$ m.

4. Design of the collecting channel
   Width of bottom $B = L \times \cos \beta°$
   $B = 0.44 \times \cos 7.5° = 0.436$ m
   For construction purposes, $B = 0.44$ m was considered.
   To calculate the pressure of the water in the collecting channel, the continuity law formula: $Q = A \times V$ and the Manning speed formula were used, whereby the equivalent expression for the flow is:

$$Q = A \times \frac{1}{n} \times R^{2/3} \times S^{1/2}$$

Known data:
$Q = 0.300$ m³/s
$n = 0.016$ Manning coefficient of the concrete, smooth finish
$B = 0.44$ m, width of the channel bed

$$R = \frac{A}{p} = \frac{0.44 \times Y}{(0.44 + 2Y)}$$

$S = 2.5\%$. This value is assumed to be the channel slope, so that the speed of the water is between 1.5 and 2.5 m/s: thus, the thick material going through the screen is carried towards the settling basin or the silt basin. Replacing data in the equivalent expression, based on Y:

$$0.300 = 0.44 \times Y \times \frac{1}{0.016} \left(\frac{0.44 \times Y}{0.44 + 2Y} \right)^{2/3} \times (0.025)^{1/2}$$

$$0.0.689 = Y \left(\frac{0.44Y}{0.44 + 2Y} \right)^{2/3}$$

To identify the value of Y, repeatedly assign values to Y until the same or approximate values are obtained in both cases.
Table A2.1 summarizes the values given to Y. The corresponding value is $Y = 0.275$ m.

Table A2.1 Results from calculations

Values of Y	0.0689
0.30	0.076
0.29	0.073
0.275	0.686

Verification of speed:

$$V = \frac{1}{0.016} \times \left(\frac{0.44 \times 0.275}{0.44 + 2 \times 0.275}\right)^{2/3} \times (0.025)^{1/2} = 2.47 \text{ m/s}$$

Comparing speeds: 2.47 m/s < 2.5 m/s ... OK.

To simplify matters, take depth t of the channel to be equivalent to the value of B and add 25 per cent more for a freeboard.
Therefore:
$t = B = 0.44$ m
Free border $= 0.25t = 0.25(0.44) = 0.11$ m
Then the total depth of the channel is: $0.44 + 0.11 = 0.55$ m.
Comparing the calculated pressure or depth of the water $Y = 0.275$ m, the assumed value of t, equivalent to B is guaranteed, more so with the larger freeboard.
Therefore, the inner dimensions of the collecting channel are conditioned to work normally at all times (Figure A2.3).
The final dimensions of the collecting ditch are shown in Figures A2.4 and A2.5.

To calculate the guide walls, use the same methodology employed in the Example 1, analysing the external forces and pressures that the stability conditions must withstand.
- The weir and the collecting channel form a single structure.
- The weir was designed of simple concrete resilient to abrasion, $f'_c = 210$ kg/cm^2 + 25% large stones up to two-thirds of its height and medium-sized stones in the top third.
- The collecting ditch was designed of reinforced concrete $f'_c = 210$ kg/cm^2, with no large or medium stones, but with sand, gravel and cement in the proportion of 1:2:2.

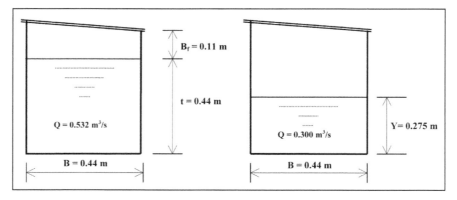

Figure A2.3 Inner dimensions of the collecting channel

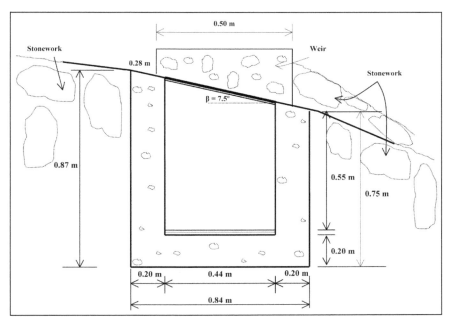

Figure A2.4 Dimensions at the beginning of the cross-section of the collecting channel. The height increases gradually, in accordance with the established slope

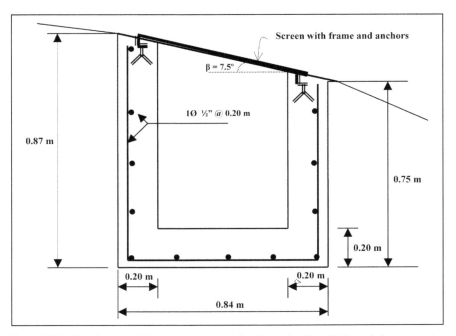

Figure A2.5 Distribution of the reinforcement in the collecting ditch and the concrete: $f'_c = 210$ kg/cm^2

EXAMPLE 3
Calculation of the spillway for the headrace channel (see Chapter 3)

The calculations made in Example 1 'Design of the intake aperture' of this Annexe determined the geometrical and hydraulic characteristics of the headrace channel for both the dry season and the rainy season. With these values, the spillway channel was designed, to be located on the side of the headrace channel, before the settling basin.

In normal conditions:
$Y_c = 0.40$ m
$Q_n = 0.625$ m³/s
$b = 0.90$ m

In high-flow conditions:
$Y'_c = 0.53$ m
$Q' = 0.887$ m³/s

Calculation of height h of the crown:
$h = Y'_c - Y_c$
$h = 0.53 - 0.40 = 0.13$ m

Calculation of the surplus flow:
$Q_e = 0.887 - 0.625 = 0.262$ m³/s

Calculation of L:
$Q = C \times L \times h^{1.5} \rightarrow 0.262 = 1.6 \times L \times 0.13^{1.5}$
Clearing $L = 3.49$ m ≈ 3.50 m
The freeboard of the spillway is assumed to be 0.15 m

The final design of this spillway and the headrace channel is shown in Figures A3.1 and A3.2.

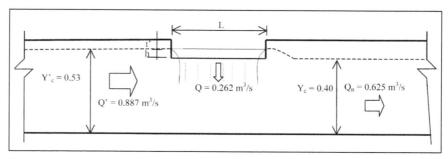

Figure A3.1 Design of the spillway

CALCULATION OF THE SPILLWAY FOR THE HEADRACE CHANNEL 283

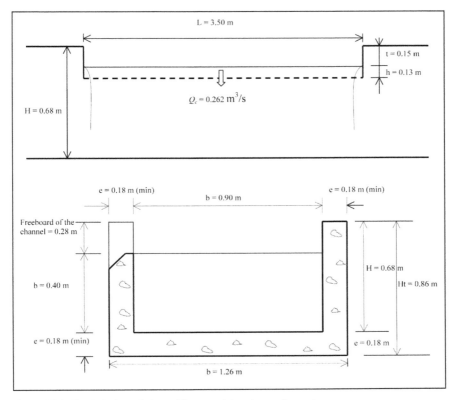

Figure A3.2 Final design of the spillway and headrace channel

EXAMPLE 4
Calculations for a coarse settling basin (see Chapter 4)

Calculating settling and collection depths, d_2 and d_1

Data:
$Q = 0.625$ m³/s
$V_h = 0.70$ m/s (assumed, less than 50 per cent of the speed in the channel)
Diameter of particles: more than 1 mm
$V_d = 0.15$ m/s (more than V_d of the 0.5 mm particle)
$f = 2$
$W = 0.90$ m (the same width as the channel)

Calculation of d_d:

From the formula: $W = \dfrac{Q}{V_h \times d_d}$, clear d_d, $d_d = \dfrac{Q}{V_h \times W}$

$d_d = \dfrac{Q}{V_h \times W} \rightarrow d_d = \dfrac{0.625}{0.7 \times 0.90} = 0.98$ m

Calculating the settling length

$L_d = \dfrac{0.7}{0.15} \times d_d \times f \rightarrow L_d = \times 0.98 \times 2.0 = 9.14\text{m} \approx 9.20\text{m}$

Calculating the total length of the settling basin

$L = L_e + L_d + L_s \rightarrow$ recommended: $L_e = L_s = W = b = 0.90$ m
$L = 0.90 + 9.20 + 0.90 = 11.00$ m

Calculating d_1 and d_2

In this case, $d_1 = d_d = 0.98$ m, as it is greater than the water pressure in the channel.

The value of d_2, is determined by adding to d_1 the difference in level resulting from the application of a 3 per cent slope to L_d.
$d_2 = 0.98 + 0.03 \times 9.20$
$d_2 = 1.256 = 1.26$ m

The final inner dimensions are shown in Figure A4.1.

CALCULATIONS FOR A COARSE SETTLING BASIN

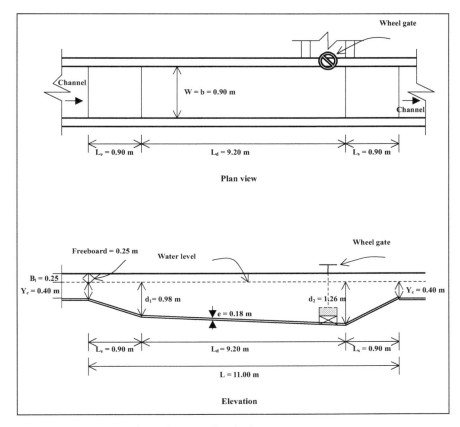

Figure A4.1 Inner dimensions of the settling basin

Thickness and steel reinforcement

For reinforced concrete structures of this kind, a minimum thickness of 0.18 m is recommended, with intercalated 3/8" (10 mm) horizontal and vertical steel reinforcements, 1/2" (13 mm) diameter; the quality of the concrete must be at least 175 kg/cm^2 (see Figure A4.2).

Calculating the dimensions of the gate

Use the outflow formula for sluice gates comprising a flat, upright, movable slab, which in this case is located sideways at the bottom of the settling basin.

This regulates the height of the orifice and the outflow. The orifice consists of width B on the floor of the settling basin and the lower border of the slab forming the gate.

286 DESIGNING MINI AND MICRO HYDROPOWER SCHEMES

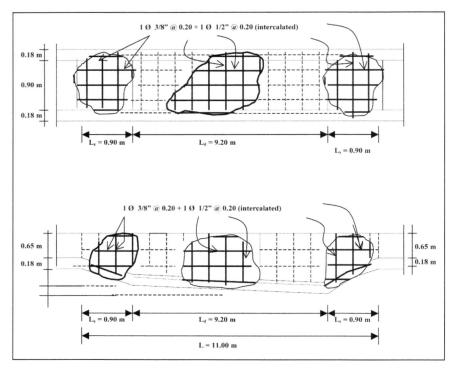

Figure A4.2 Distribution of the steel structure in the settling basin

The formula will reveal the dimensions of the (rectangular) orifice, so that the dimensions of the sluice opening and the gate can be determined for construction purposes.

Data on the above example:
$k = 1$ (free discharge)
$\mu = 0.60$
$a = 0.030$ m (assumed, based on the particles to be permanently expelled: 1 mm $< \emptyset <$ 20 mm)
$B = 0.25$ m (assumed, for a first trial)
$y_1 = 1.12$ m (average of d_1 and d_2)

Replace these values in Q

$Q = k \times \mu \times a \times B \times \sqrt{2 \times g \times y_1}$

$Q = 1 \times 0.60 \times 0.030 \times 0.25 \times \sqrt{2 \times 9.8 \times 1.12}$

$Q = 0.021$ m³/s

This permanent outflow is considered in the intake flow design (0.625 m³/s):
$a = 0.030$ m (opening)
$B = 0.25$ m

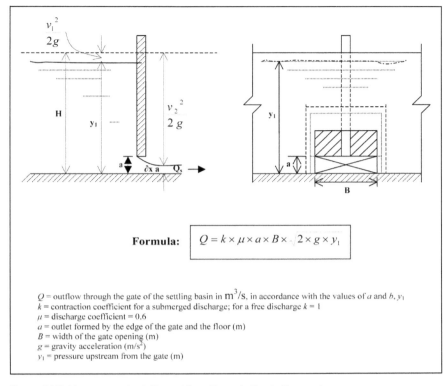

Figure A4.3 Measurement of the outflow through the bottom gate

For maintenance purposes, the gate will measure 0.25 m × 0.20 m. A wheel gate was chosen because of its regulation and maintenance characteristics.

EXAMPLE 5
Calculations for an open channel (see Chapter 5)

The following field data were used to calculate the dimensions of the channel for three kinds of open section with a maximum hydraulic efficiency (rectangular, trapezoidal and semicircular):

Length of channel = 495 m
Design slope = 2 per thousand
Concrete material, polished finish, Manning coefficient n = 0.013
Design flow: 0.600 m³/s

For each case, the calculations are made in an iterative manner with the help of Tables A5.1–A5.3, depending on the hydraulic conditions, until the assumed data corresponded to the design flow.

For a rectangular section

Table A5.1 Parameters to be calculated

b (m)	y(m)	A(m²)	P(m)	R(m)	n	S	1/n	$R^{2/3}$	$S^{1/2}$	V(m/s)	Qm³/s

b = assumed value
$y = b/2$, for a maximum hydraulic efficiency section
A, P, R = calculated with the area of the rectangle, applying the respective ratios
n = information obtained from the Manning coefficient table, in accordance with the channel material
S = design slope provided by the project designer
$1/n$, $R^{2/3}$ and $S^{½}$ were processed with a calculator
V = is calculated using the corresponding formula $V = \dfrac{1}{n} \times R^{2/3} \times S^{1/2}$
Q = is calculated using the continuity formula: $Q = A \times V$

For a trapezoidal section

Table A5.2 Parameters to be calculated

b(m)	y(m)	T(m)	A(m²)	P(m)	R(m)	n	S	1/n	$R^{2/3}$	$S^{1/2}$	V(m/s)	Qm³/s

b = assumed value
$y = 0.865\,b$, for a maximum hydraulic efficiency section, when $\theta = 60°$
$T = 2b$, for a maximum hydraulic efficiency section
A, P and R are calculated with the formula, the area of the trapezium and the indicated ratios
n = roughness factor, depending on the channel material (table of Manning coefficients)
$1/n$, $R^{2/3}$ and $S^{1/2}$ are processed with a calculator

V = is calculated using the corresponding formula $V = \dfrac{1}{n} \times R^{2/3} \times S^{1/2}$

Q = is calculated using the continuity formula: $Q = A \times V$

For a semicircular section

Table A5.3 Parameters to be calculated

y(m)	T=D(m)	A(m²)	P(m)	R(m)	S	n	1/n	R^{2/3}	S^{1/2}	V(m/s)	Q(m³/s)

y = assumed value
$D = 2y$, for a maximum hydraulic efficiency section
$T = D$, for a maximum hydraulic efficiency section
A, P and R are calculated with the area of the semi-circle, semi-circumference and the established ratios
n = roughness factor, depending on the channel material (table of Manning coefficients)
$1/n$, $R^{2/3}$ and $S^{1/2}$, are processed with a calculator

V = is calculated using the corresponding formula $V = \dfrac{1}{n} \times R^{2/3} \times S^{1/2}$

Q = is calculated using the continuity formula $Q = A \times V$

Based on these recommendations, the calculations and the recorded figures appear in Figure A5.1.

Adding the freeboard, thickness and the inside and outside crown to these sections, the following can be estimated:
- The total width of the platform required on the site.
- How much the slope must be sheared to form the platform.
- The size of the excavation, etc. and the type of section that would involve the least investment.

'Referential' criteria for determining the freeboard:
1. Based on the pressure: $B_f = 0.25y$ to $0.33y$
2. Based on the flow (see Table A5.4):

290 DESIGNING MINI AND MICRO HYDROPOWER SCHEMES

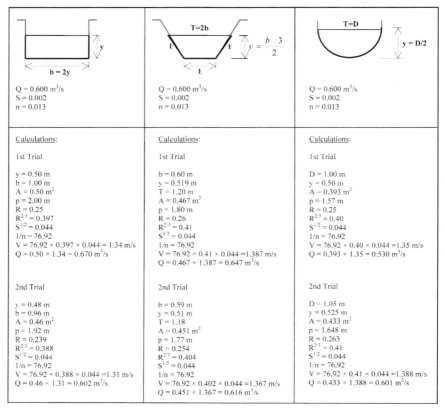

Figure A5.1 Calculation of a conveyance channel with open sections of a maximum hydraulic efficiency, for a 0.600 m³/s flow, a 0.002 slope and a Manning factor $n = 0.013$ (polished concrete casing)

Table A5.4 Recommended values for B_f according to flow

Q (m³/s)	B_f (m)
Less than 0.50	0.10 to 0.15
More than 0.50	More than 0.15

The thickness of the channel will depend on the type of soil, the total height of the channel, the angle formed by the channel walls and the horizontal, the casing method selected, the thrust of the earth on the slopes, etc.

Table A5.5 shows thicknesses for reference purposes, based on previous work.

CALCULATING AN OPEN CHANNEL

Table A5.5 Thickness of simple concrete channels

Casing method	Thickness (m)	Q (m³/s)	Section	Type of soil
Formwork	0.12–0.16	≤ 1.50	Rectangular Trapezoidal	Compact clay or rocky
Wooden frames	0.05–0.10	≤ 1.00	Trapezoidal Semicircular	Compact clay or rocky

If some sections are of reinforced concrete, then the reinforced concrete technology must be followed.

Furthermore, it is recommended that the inner crown should be at least 0.50 m, to prevent clumps of mud or small loose stones from slipping straight into the channel and affecting the normal operation of the plant.

The width of the outer crown will depend on the intended function. It is usually used during the construction of the channel to transport materials and tools and to allow the resident engineer, the supervisor, etc. to work in it. When the works are completed, it is used as a path for maintenance work and to gain access to the intake weir, the settling basin, the forebay tank, concrete structures, etc. At least 1.00 m is recommended.

Comparing the calculations made in Figure A5.1 (2nd trial), the results of Figure A5.2 and Table A5.6, the best economic alternative can be chosen, depending on local construction costs.

Table A5.6 Some measurements of the calculated sections as indices for determining the most economic section of the channel to deliver 0.600 m³/s

Section characteristics	Rectangular section	Trapezoidal section	Semicircular section
Platform (m)	2.46	2.82	2.55
Excavation of the channel ditch (m³/m)	0.864	0.873	0.808
Casing (m³/m)	0.276	0.272	0.250
Expansion joint, 0.12 m wide × 0.025 m thick	2.40	2.15	2.08

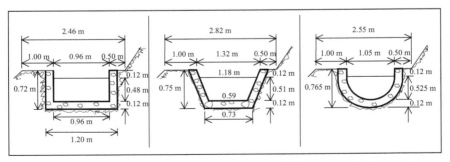

Figure A5.2 Final dimensions of the rectangular, trapezoidal and semicircular sections of a channel, to deliver 0.600 m³/s

EXAMPLE 6
Calculation of an enclosed channel (pipe channel)

Applying the same data used for the conveyance channel and the manufacturer's conditions, calculate the inner diameter:

$Q = 0.600 \text{ m}^3/\text{s}$
$S = 0.002$
$n = 0.01$
$D_i = ?$

From the formula:

$Q = 28.4 \times D_i^{8/3} \times S^{1/2}$, clear D_i

$$D_i = \left(\frac{Q}{28.4 \times S^{1/2}} \right)^{3/8}$$

$$D_i = \left(\frac{0.600}{28.4 \times 0.002^{1/2}} \right)^{3/8}$$

$D_i = 0.754 \text{ m} = 29.71" = 754 \text{ mm}$

EXAMPLE 7
Calculations for the silt basin (see Chapter 6)

Use the following field data for calculating the dimensions of the settling basin and complementary works:
- Conveyance channel
  $Q = 0.600$ m³/s
  Rectangular section: $b = B = 0.96$ m
  Head, $y = 0.48$ m
  Freeboard $B_f = 0.22$ m
  Slope $S = 0.002$
- Settling basin
  Soil: Split rock, easy to excavate with hand tools
  Platform: 7 m wide and 15 m long
  Slope 18°
  Centre line: Straight section with respect to the centre line of the channel
  Material: Reinforced concrete, 85 per cent embedded in the soil
  Cleaning system: Side wheel gate, 0.40 m × 0.40 m

1. Calculating width W and d_d

   Use the formula: $W = \dfrac{Q}{V_h \times d_d}$

   Assuming: $V_h = 0.3$ m/s
   Replacing values: $d_d = y + 0.05$ m $= 0.48 + 0.05 = 0.53$ m

   $W = \dfrac{0.6}{0.3 \times 0.53} = 3.77$ m

2. Calculating L_d
   The recommendations of the turbine manufacturer indicate that the water should not carry any particles with a diameter larger than 0.5 mm. As a precaution, it was decided that the particles should be no larger than 0.3 mm. With this value, it was observed that $V_d = 0.03$ m/s, as shown in Table 6.1 of the text (Chapter 6, Silt basin).
   Replacing values and adding safety factor $f = 2$, the following is obtained:

   $L_d = \dfrac{V_h}{V_d} \times d_d \times f \rightarrow L_d = \dfrac{0.3}{0.03} \times 0.53 \times 2 = 10.60$

3. Calculating L_e and L_s
   It is recommended that $L_e = L_s = W = 3.77$ m

4. Calculating d_1 and d_2
   Assuming: $d_1 = 1.6y$, $d_1 = 1.6 \times 0.48 = 0.79$ m
   $d_2 = d_1 + 0.04 L_d \rightarrow d_2 = 0.79 + 0.04 \times 10.60 = 1.21$ m
   $d_r = d_2 - d_1 = 1.21 - 0.79 = 0.42$ m
5. Thickness of the floor and walls:
   According to reinforced concrete recommendations, given the soil conditions and the fact that these hydraulic structures will be embedded in the ground, the floor and walls should be at least 0.20 m thick.
6. Calculation of the spillway
   In: $Q = 1.6 L \times h^{3/2}$
   Assuming $h = 0.18$ m (less than B_f)
   Q was not considered as a surplus, but as a total: $Q_{total} = 0.600$ m³/s, for safety reasons, in case the sluice gate fails to work or in the event of an emergency.

$$L = \frac{Q}{1.6 \times h^{3/2}} \rightarrow L = \frac{0.600}{1.6 \times 0.18^{3/2}} = 4.91 \text{m}$$

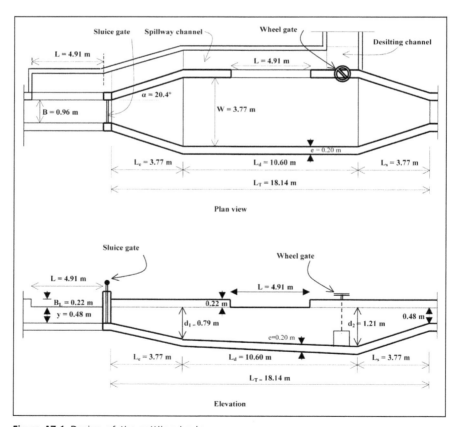

Figure A7.1 Design of the settling basin

CALCULATIONS FOR THE SILT BASIN

The final design of the settling basin and its complementary works can be seen in Figure A7.1

A spreadsheet can be used for the calculations, as shown in Table 7.2 of the text (Chapter 7, Forebay tank).

7. Design of complementary works

This consists of determining the elements of the spillway channel, applying channel formulae and known data:

$Q = 0.600$ m^3/s
$S = 0.006$ (assumed)
$b = 0.80$ m
$y = 0.40$ m (assumed for a rectangular maximum hydraulic efficiency channel)
$n = 0.014$ (concrete casing, polished finish)

With this data and the application of the formulae, verify whether the flow to be conveyed is 0.605 m^3/s at a speed of 1.89 m/s.

It is worth noting that the conveyance channel also has a spillway before the settling basin. This spillway is the same size as the settling basin spillway and is used to carry the water in the overflow channel when the sluice gate is shut to evacuate the sediments and clean the settling basin.

For the final design of the spillway channel, a 0.12 m freeboard was added to the thickness of the floor and to the 0.12 m wall, as well as two 0.25 m and 0.50 m steps to connect it to the headrace channel at a higher level. The size of this channel may vary, depending on the topography of the land.

The final design of this complementary channel is illustrated in Figure A 7.2.

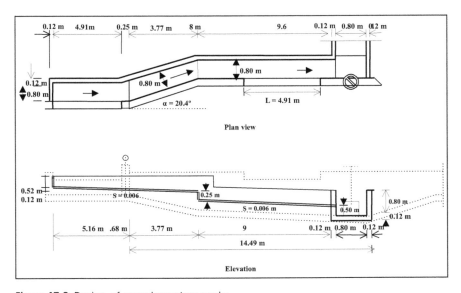

Figure A7.2 Design of complementary works

The steel reinforcement in the settling basin is distributed as shown in Figure A7.3.

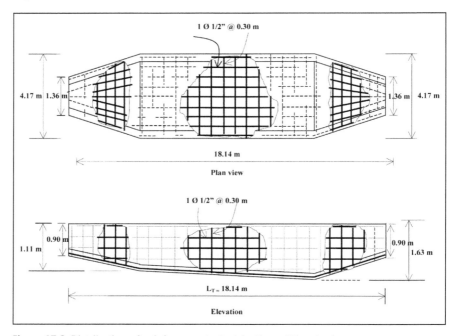

Figure A7.3 Distribution of reinforcement steel in the settling basin

EXAMPLE 8
Calculations for the forebay tank (see Chapter 7)

To determine the dimensions of the forebay tank, use the same values that were employed to determine the dimensions of the settling basin.
1. Calculate the length = L_s of the settling basin = 3.77 m
2. Calculate the width = W of the settling basin = 3.77 m
3. Calculate the depth $d_3 = 4.5D + d$

Determine the inner diameter of the pressure pipe $D_i = \sqrt{\dfrac{4 \times Q}{\pi \times V}}$

Q = flow in m³/s
V = speed of the water in the pipe (m/s)
For a first reliable approximation, it is assumed that $V = 2$ m/s
Replacing values $D_i = 0.62$ m
$d = 0.50$ m (assumed)
Then: $d_3 = 4.5(0.62) + 0.50 = 3.29$ m
Depth of the bottom from the roof to the floor = $d_3 + B_f = 3.29 + 0.22 = 3.51$ m

4. Height of the wall supporting the screen (with respect to the floor of the forebay tank) = $d_3 - d_d = 3.29 - 0.53 = 2.76$ m
5. Thickness of the floor and the forebay tank = 0.25 m (Table 14.2)

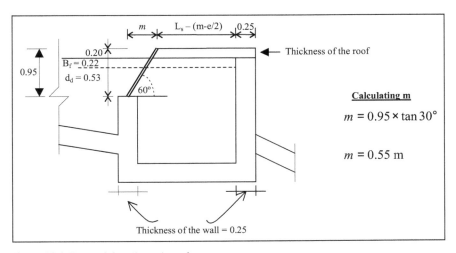

Figure A8.1 Determining the value of m

6. Calculate the dimensions of the roof:
   Width = $W + 2e = 3.77 + 2 \times 0.25 = 4.27$ m
   Length = $L_s + 1.5e - m$
7. Determine the value of m in Figure A8.1.
   Length = $3.77 + 1.5 (0.25) - 0.55 = 3.595$
   Thickness of the roof = 0.20 (Table 7.4, Chapter 7, Forebay tank)

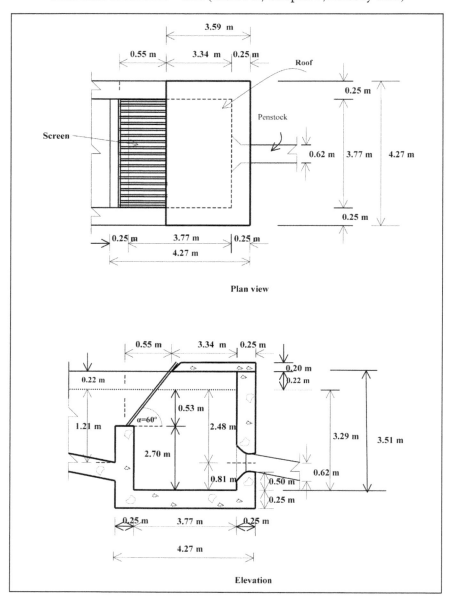

Figure A8.2 Final sizing of the forebay tank components joined to the settling basin

CALCULATIONS FOR THE FOREBAY TANK

8. Dimensions of the screen:
   Length = $W - 1$ cm = $3.77 - 0.01 = 3.76$ m
   Width = $[(d_d + B_L + \text{thickness of the roof})^2 + m^2)]^{0.5}$
   = $[0.95^2 + 0.55^2]^{0.5} = 1.097$ m ≈ 1.10 m

Figures A8.2 and A8.3 show the calculated dimensions and details of the safety screen. Figure A8.4 shows the steel distribution in the floor, walls and roof of the forebay tank.

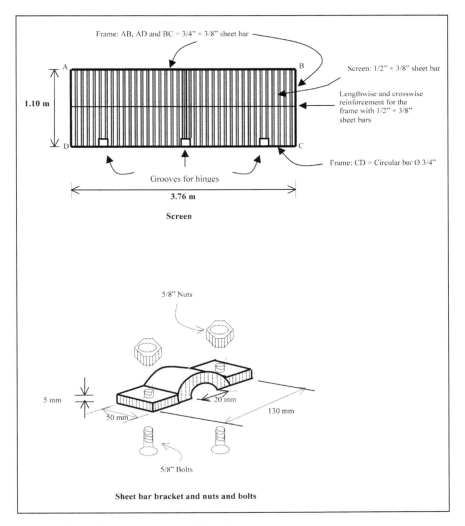

Figure A8.3 Details of the screen and bracket
Note: 3/8" = 10 mm; ½" = 13 mm; 5/8" = 16 mm; ¾" = 19 mm

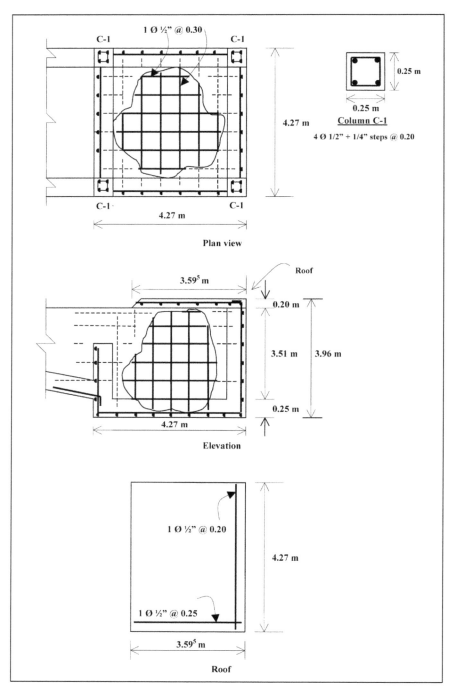

Figure A8.4 Steel distribution in the forebay tank
Note: ¼" = 6 mm; ½" = 13 mm

EXAMPLE 9
Calculations for the PVC penstock (see Chapter 8)

Field data

- Using a clinometer and a measuring tape, the following distances and angles were noted and a sketch was drawn of the ground profile at the site of the penstock in the Sondor MHS (120 kW) (Figure A9.1).
- The route of the penstock was selected as the most adequate and geologically stable, considering other aspects:
  o   The short distance.
  o   A single alignment, at right angles with the wall in which the forebay tank would be embedded.
  o   Entry to the forebay tank should not be too forced with respect to the vertical plan.
  o   The higher part is clear of any rocks or large stones that could be a potential hazard for the powerhouse.
  o   The facility to build a path from the powerhouse to the forebay tank for operational purposes, which also provides access for the installation of the piping.
  o   Advantage must be taken of the relief of the ground and adjacent natural courses to control the drainage of rain water, or reduce the costs of complementary drainage works.
  o   The owner of the land must be assured that the piping will not affect the farmland, as the pipes will be buried at least 0.80 m underground.

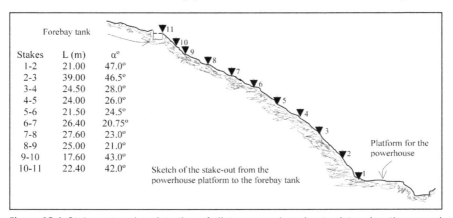

Figure A9.1 Stake out and registration of distances and angles to determine the ground profile

302 DESIGNING MINI AND MICRO HYDROPOWER SCHEMES

Data processing

- It was determined that FU PVC pipes would be used, as pipes capable of carrying 150 m of pressure were locally available.
- The ground profile and levels were determined: Figure A9.2 (a) and (b).
- The profile of the penstock was defined, in order to obtain more important data: number of sections, the length of each section, the pressure head in each section, etc. (Figure A9.2 (c)).

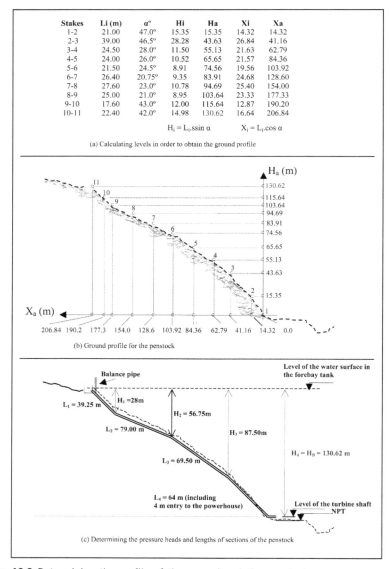

Stakes	Li (m)	$a°$	Hi	Ha	Xi	Xa
1-2	21.00	47.0°	15.35	15.35	14.32	14.32
2-3	39.00	46.5°	28.28	43.63	26.84	41.16
3-4	24.50	28.0°	11.50	55.13	21.63	62.79
4-5	24.00	26.0°	10.52	65.65	21.57	84.36
5-6	21.50	24.5°	8.91	74.56	19.56	103.92
6-7	26.40	20.75°	9.35	83.91	24.68	128.60
7-8	27.60	23.0°	10.78	94.69	25.40	154.00
8-9	25.00	21.0°	8.95	103.64	23.33	177.33
9-10	17.60	43.0°	12.00	115.64	12.87	190.20
10-11	22.40	42.0°	14.98	130.62	16.64	206.84

$H_i = L_i \cdot \sin \alpha$ $X_i = L_i \cdot \cos \alpha$

(a) Calculating levels in order to obtain the ground profile

(b) Ground profile for the penstock

(c) Determining the pressure heads and lengths of sections of the penstock

Figure A9.2 Determining the profile of the ground and the penstock

In Figure A9.2 (a):
- H_i = the partial head = $L_i \times \sin \alpha$
- H_a = the accumulated head
- H_B = the gross head, equivalent to the last accumulated head
- X_i = partial horizontal distance = $L_i \times \cos \alpha$
- X_a = accumulated horizontal distance

In Figure A9.2 (b):
- At an adequate scale, the points (ground levels) are identified from 1 to 11, determined by the coordinates (X_a, H_a).
- Then these points are joined, in accordance with the sketch drawn in the field, and the ground profile is determined.

In Figure A9.2 (c):
- Based on the ground profile, determine the profile of the penstock by drawing straight sections of piping from the forebay tank downhill, or from the powerhouse uphill. The length of each section depends on where the piping must change direction due to the topography of the land. Efforts must be made to change direction as few times as possible, bearing in mind that the PVC penstock will be buried at least 0.80 m, and excessive digging must be prevented.
- The location of the forebay tank and the water level must be drawn to scale before starting from the forebay tank. Then the initial point of the penstock must be located at a depth equivalent to four times the diameter from the water surface.
- To start from the powerhouse, the turbine manufacturer must be consulted about the level at which the penstock must 'enter horizontally' with respect to the finished floor of the powerhouse.
- Try to ensure that the profile drawn to scale has no precision errors, as this profile will determine the pressure heads, the lengths of each section and the angles they form with each other and with the horizontal, for the corresponding calculations.

Design data and desk work

- $Q = 0.150$ m³/s (design flow for the turbine)
- $H_B = 130.62$ m
- 1st section of the pipe: $L_1 = 39.25$ m, $H_1 = 28.00$ m
- 2nd section of the pipe: $L_2 = 79.00$ m, $H_2 = 56.75$ m
- 3rd section of the pipe: $L_3 = 69.50$ m, $H_3 = 87.50$ m
- 4th section of the pipe: $L_4 = 64.00$ m, $H_4 = 130.62$ m

Note: Length L_4 which enters the powerhouse includes 4 m of horizontal piping to be coupled to the turbine.

- Total length of the pipe $L_T = L_i = 251.75$ m
- Number of changes of section = 4 (each section of the PVC pipe has the same outer diameter; the inner diameter changes slightly due to the different thickness of the wall)
- Number of changes of direction = 4

(Tentative) calculation of the inner diameter for each section

$$d_i = \sqrt{\frac{4 \times Q}{\pi \times V}} \text{ assuming } V = 2.2 \text{ m/s} \rightarrow d_i = \sqrt{\frac{4 \times 0.150}{3.14 \times 2.2}} = 0.294\text{m}$$

As the chosen material is flexible union PVC, from the corresponding catalogue (see Table 8.3) we find that pipes are available with an outer diameter of 315 mm = 0.315 m, class 5, class 7.5, class 10 and class 15.

It is indicated that class 5 pipes have a capacity for 50 m head, class 7.5 up to 75 m head, class 10 up to 100 m head and class 15 up to 150 m head.

Consequently, from Table 8.3, outer diameter d_e and thickness t were noted and inner diameter d_i was deducted to 'verify' the choice:

Table A9.1 Matching head of penstock to commercial PVC pipes according to class Processed data (from Figure A9.2 (c)) Catalogue data

	Processed data			Catalogue data		
Section	Length L_i (m)	Pressure head H_i (m)	Class	d_e (m)	t (mm)	d_i (m)
1	39.25	28.00	5	0.315	7.7	0.299
2	79.00	56.75	7.5	0.315	11.4	0.292
3	69.50	87.50	10	0.315	15	0.285
4	64.00	130.62	15	0.315	22	0.271

Head loss due to friction

Calculating head loss due to friction, h_p, (in each section of the pipe):

$$h_p = h_f + h_t$$

a) $h_f = 0.08 \times \dfrac{f \times L_i \times Q^2}{d_i^5}$

b) $h_t = \dfrac{V^2}{2g}(K_1 + K_2 + \ldots K_n)$

Calculating f:

This is obtained from the Moody diagram, but first find the values of:

$1.27 \times \dfrac{Q}{d_i}$ and K/d_i

in each section of the pipe.

Table A9.2 Data processing for different sections of the penstock

Section	Q (m³/s)	d_i (m)	1.27Q/d_i (m³/s)/m	K (mm)	K/d_i	f
1	0.150	0.299	0.637	0.01	0.0000334	0.0138
2	0.150	0.292	0.652	0.01	0.0000342	0.0136
3	0.150	0.285	0.668	0.01	0.0000351	0.0134
4	0.150	0.271	0.703	0.01	0.0000369	0.0132

Note: To obtain the values of column K/d_i, both values are expressed in the same unit

Once the value of f in each section is known, calculate the loss of head or load loss due to friction:

$$h_f = 0.08 \times \frac{f \times L_i \times Q^2}{d_i^5}$$

Table A9.3 Calculation of friction losses for different sections

Section	f	L_i (m)	Q (m³/s)	d_i (m)	h_f (m)
1	0.0138	39.25	0.150	0.299	0.41
2	0.0136	79.00	0.150	0.292	0.91
3	0.0135	69.50	0.150	0.285	0.90
4	0.0132	64.00	0.150	0.271	1.04

Head loss due to turbulence

To calculate the head loss due to turbulence in each section of the pipe, consider factor K (Table 8.1) for the inlet, curves, change of section, valves and other accessories.

$$h_t = \frac{V^2}{2g}(K_1 + K_2 + ...K_n)$$

The speed in each section is calculated with the formula: $V = \dfrac{4Q}{\pi \times d_i^2}$

Table A9.4 Calculations of speed for different sections of penstock

Section	d_i (m)	V (m/s)	$V^2/2g$	K = r/d	($K_1+K_2+...$)	h_t
1	0.299	2.13	0.233	K_{inlet} = 0.50 K_{curve} = 0.25	0.75	0.1750
2	0.292	2.23	0.256	K_{curve} = 0.25	0.25	0.064
3	0.285	2.35	0.282	K_{curve} = 0.25	0.25	0.071
4	0.271	2.60	0.350	K_{curve} = 0.38 $K_{gate\ valve}$ = 0.10	0.48	0.168

This step can be omitted when the calculated values are lower than the loss by friction values.

Therefore, the *head loss due to friction and turbulence in each section is*:

$h_p = h_f + h_t$

Table A9.5 Calculation of total losses for different sections of the penstock

Section	h_f (m)	h_t (m)	$h_p = h_f + h_t$ (m)
1	0.41	0.18	0.59
2	0.91	0.06	0.97
3	0.90	0.07	0.97
4	1.06	0.17	1.23

Calculating the load loss by friction, in percentage terms

$$Losses = \frac{h_p}{H_B} \times 100$$

Table A9.6 Losses as a percentage of the head

Section	h_p(m)	H_B(m)	% of losses due to friction and turbulence
1	0.59	130.62	0.45
2	0.97	130.62	0.74
3	0.97	130.62	0.74
4	1.23	130.62	0.94
			Σ% Losses = 2.87%

The total loss shown in Table A9.6 is small, therefore it is considered acceptable. As discussed in Chapter 8 the percentage of losses depends on the criteria of the designer and the site conditions, but generally is below 10% of the total head.

Verifying the selected wall thickness of the penstock

a) First, in each section of the piping, calculate wave propagation speed a caused by the valve being shut.

$$a = \frac{1420}{\sqrt{1 + \left(\frac{E_{ag} \times d_i}{E_{pvc} \times t}\right)}}$$

CALCULATIONS FOR THE PVC PENSTOCK

Table A9.7 Calculation of values for a

Section	E_{AG} (kg-f/cm²)	d_i (m)	$E_{AG} \times d_i$	E_{PVC} (kg-f/cm²)	t (m)	$E_{PVC} \times t$	a (m/s)
1	21,000	0.299	6279	28,000	0.0077	215.6	258.72
2	21,000	0.292	6132	28,000	0.0114	319.2	315.86
3	21,000	0.285	5985	28,000	0.015	420.0	363.62
4	21,000	0.271	5691	28,000	0.022	616.0	443.78

b) Having determined the wave speed in each section, determine the highest pressures Δh created in each section of the piping.

$$\Delta h = \frac{a \times V}{g}$$

Table A9.8 Pressure increase due to surge

Section	a (m/s)	ΔV (m/s)	$a \times V$ (m²/s²)	g (m/s²)	Δh (m)
1	258.72	2.13	551.07	9.8	56.22
2	315.86	2.23	704.37	9.8	71.87
3	363.62	2.35	854.51	9.8	87.19
4	443.78	2.60	1153.83	9.8	117.73

c) To check whether the selected pipe thickness in each section is correct, it should have the capacity to support a maximum head pressure, h_T (losses by friction are not taken into account as they are minor):

$$h_T = \Delta h + H_i$$

Table A9.9 Total pressure head including surge pressure

Section	Δh (m)	H_i (m)	$h_T = \Delta h + h_p$ (m)
1	56.22	28.00	84.22
2	71.87	56.75	128.62
3	87.19	87.50	174.69
4	117.73	130.60	248.33

d) Calculating the safety factor in each section of the pipe:

$$f_s = \frac{t \times \sigma}{5 \times h_T \times d_i \times 10}$$

Table A9.10 Calculation of the safety factor for each section of the penstock

Section	t(mm)	σ_{PVC} (kg/cm²)	h_T (m)	d_i(m)	f_s
1	7.71	500	84.22	0.299	3.06
2	11.4	500	128.62	0.292	3.03
3	15	500	174.69	0.285	3.01
4	22	500	248.33	0.271	3.27

As recommended in Chapter 8, the minimum safety factor in each section is 3, therefore it is certified that the selected diameter and thickness of the pipes are correct for installation purposes.

EXAMPLE 10

Design of the RU PVC and steel pipe, Chetilla MHS, Cajamarca, Peru (see Chapter 8)

Data

$Q = 0.080 \text{ m}^3/\text{s}$
$H_B = 175.00 \text{ m}$
Turbine = horizontal shaft Pelton with two jets

In accordance with the field data, the following profiles of the ground and the penstock were obtained from desk work (Figure A10.1).

- 1st section of the pipe: $L_1 = 60.00 \text{ m}$, $H_1 = 34.30 \text{ m}$
- 2nd section of the pipe: $L_2 = 60.00 \text{ m}$, $H_2 = 69.00 \text{ m}$
- 3rd section of the pipe: $L_3 = 74.00 \text{ m}$, $H_3 = 121.50 \text{ m}$
- 4th section of the pipe: $L_4 = 105.20 \text{ m}$, $H_4 = 175.00 \text{ m}$

Length L_4 entering the powerhouse includes 3.20 m of horizontal pipes to be coupled to the turbine.

- Total length of the pipe $L_T = L_i = 299.20\text{m}$
- Number of section changes = 4

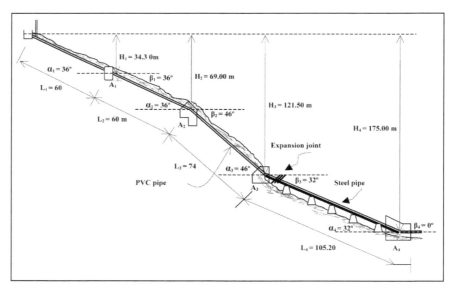

Table A10.1 Profile of the penstock with steel pipes in the lower part and PVC pipes in the higher part

DESIGN OF THE RU PVC AND STEEL PIPE

(Tentative) calculation of the inner diameter of the pipe for each section

$$d_i = \sqrt{\frac{4 \times Q}{\pi \times V}}, \text{ assuming } V = 2.5 \text{ m/s} \to d_i = \sqrt{\frac{4 \times 0.080}{3.14 \times 2.5}} = 0.201 \text{ m}$$

It was determined that the pipe for the fourth section (lower part) would be steel, schedule 40, to withstand the high pressure (H_B) of 175 m, plus the transitory pressure caused by the valve being shut; and RU PVC material for the 3rd, 2nd and 1st sections, to endure the lower pressure of 150 m and also because metal accessories for joining steel and PVC pipes are available in the local market.

From the catalogue (Table 8.7), the data in Table A10.1 were obtained:

Table A10.1 Data for commercial PVC pipes

Class	d_e (m)	t (mm)	d_i (m)
7.5	0.219	7.9	0.203
10	0.219	10.4	0.198
15	0.219	15.3	0.188

and from the catalogue (Table 8.8) for the steel pipe:

Table A10.2 Data for commercial steel pipes

Schedule	d_e(m)	t(mm)	d_i(m)
40	0.219	8.178	0.202

Table A10.3 contains a summary of the data to be followed to verify the selected diameter and thickness.

Table A10.3 Results for the different materials to be used for the different sections of the penstock

	Processed data (from Figure A10.1)		Catalogue data		
Section	Length L_i (m)	Pressure head H_i (m)	Class or schedule	d_e (m)	t(mm)	d_i m
1	60	34.30	7.5 PVC	0.219	7.9	0.203
2	60	69.00	10 PVC	0.219	10.4	0.198
3	74	121.50	15 PVC	0.219	15.3	0.188
4	105.2	175.00	40 steel	0.219	8.178	0.202

Calculating the total load loss h_p, (in each section of the pipe)

$h_p = h_f + h_t$

a) $h_f = 0.08 \times \dfrac{f \times L_i \, Q^2}{d_i^5}$

b) $h_t = \dfrac{V^2}{2g} (K_1 + K_2 + \ldots K_n)$

Calculating f
The following values were obtained for each section:

$1.27 \times \dfrac{Q}{d_i}$, and K/d_i

With these values, the Moody diagram was consulted to determine the value of f. The corresponding calculations appear in Table A10.4.

Table A10.4 Calculating f

Section	Q (m³/s)	d_i (m)	$1.27 Q/d_i$ (m³/s)/m	K (mm)	K/d_i	f
1	0.080	0.203	0.50049	0.01 PVC	0.0000492	0.014
2	0.080	0.198	0.51313	0.01 PVC	0.0000505	0.014
3	0.080	0.188	0.54042	0.01 PVC	0.0000531	0.0138
4	0.080	0.202	0.50297	0.1 steel	0.0004950	0.0178

Note: The K/d_i ratio should be expressed in the same unit.

Replacing values, calculate the head or load loss due to friction

a) $h_f = 0.08 \times \dfrac{f \times L_i \, Q^2}{d_i^5}$

Table A10.5 Calculation of friction losses

Section	f	L_i (m)	Q (m³/s)	d_i (m)	h_f (m)
1	0.014	60	0.080	0.203	1.24
2	0.014	60	0.080	0.198	1.41
3	0.0138	74	0.080	0.188	2.23
4	0.0178	105.2	0.080	0.202	2.85

b) Calculating the head loss due to turbulence in each section of the pipe

$h_t = \dfrac{V^2}{2g} (K_1 + K_2 + \ldots K_n)$

The speed in each section is calculated with the formula: $V = \dfrac{4Q}{\pi \times d_i^2}$

DESIGN OF THE RU PVC AND STEEL PIPE

Table A10.6 Summary of the calculations

Section	d_i (m)	V (m/s)	$V^2/2g$	$K = r/d$	$(K_1+K_2+...)$	H_t
1	0.203	2.47	0.311	$K_{inlet} = 0.50$	0.50	0.16
2	0.198	2.59	0.342	$K_{contrac} = 0.20$	0.45	0.15
				$K_{curve} = 0.25$		
3	0.188	2.87	0.420	$K_{contrac} = 0.20$	0.45	0.19
				$K_{curve} = 0.25$		
4	0.202	2.48	0.314	$K_{curve} = 0.38$	0.48	0.15
				$K_{gate\ valve} = 0.10$		

Therefore, the head loss due to friction and turbulence in each section is:
$h_p = h_f + h_t$

Table A10.7 Total losses in the penstock

Section	h_f(m)	h_t(m)	$h_p = h_f + h_t$(m)
1	1.24	0.16	1.40
2	1.41	0.15	1.56
3	2.23	0.19	2.42
4	2.85	0.15	3.00

Calculating the load loss by friction, in percentage terms

$$Losses = \frac{h_p}{H_B} \times 100$$

Table A10.8 Losses as a percentage of the pressure head

Section	h_p (m)	H_B (m)	% Losses due to friction and turbulence
1	1.40	175.00	0.008
2	1.56	175.00	0.009
3	2.42	175.00	0.014
4	3.00	175.00	0.017
			$\Sigma\% = 4.6\%$

4.6 % < 10% ... OK

Verifying the wall thickness of the penstock

a) Calculate wave propagation speed, a, caused by the valve being shut.

$$a = \frac{1420}{\sqrt{1 + \left(\frac{E_{ag} \times d_i}{E_{pvc} \times t}\right)}}$$

312 DESIGNING MINI AND MICRO HYDROPOWER SCHEMES

Table A10.9 Calculation speed of the propagation wave

Section	E_{AG} (kg-f/cm²)	d_i (m)	$E_{AG} \times d_i$	E (kg-f/cm²)	t (m)	Ext	Δh (m/s)
1	21,000	0.203	4263	28,000 PVC	0.0079	221.2	315.38
2	21,000	0.198	4158	28,000 PVC	0.0104	291.2	363.28
3	21,000	0.188	3948	28,000 PVC	0.0153	428.40	444.27
4	21,000	0.202	4242	2,000,000 STEEL	0.0082	16,400	1265.71

b) With these values of a, calculate the highest pressures Δh created in each section of the pipeline.

$$\Delta h = \frac{a \times V}{g}$$

Table A10.10 Pressure in each section of the penstock

Section	Δh (m/s)	V (m/s)	$\Delta h \times V$ (m²/s²)	g (m/s²)	Δh (m)
1	315.38	2.47	778.99	9.8	79.49
2	363.28	2.59	840.89	9.8	96.00
3	444.27	2.87	1275.05	9.8	130.00
4	1265.71	2.48	3138.96	9.8	320.30

c) Verification of the selected pipe thickness in each section:
The thickness should endure both the maximum transitory pressure and the gross head, in each section of the pipeline.

$$h_T = \Delta h + H_i$$

Table A10.11 Total loss in each section of the penstock

Section	Δh (m)	H_i (m)	$H_i = \Delta h + h_p$ (m)
1	79.49	34.30	113.79
2	96.00	69.00	165.00
3	130.00	121.50	251.50
4	320.30	175.00	495.30

d) Calculating the safety factor in each section of the pipeline

For the PVC sections: $f_s = \dfrac{t \times \sigma}{5 \times h_T \times d_i \times 10}$

For the steel section: $f_s = \dfrac{t \times \sigma}{5 \times h_T \times d_i \times K_j \times 10}$

DESIGN OF THE RU PVC AND STEEL PIPE 313

Table A10.12 Materials and their characteristics for each section of the penstock

Section	t (mm)	σ (kg/cm^2)	h_T (m)	d_i (m)	f_s
1 (PVC)	7.9	500 PVC	113.79	0.203	3.42
2 (PVC)	10.4	500 PVC	165.00	0.198	3.18
3 (PVC)	15.3	500 PVC	251.50	0.188	3.23
4 (Steel), $K_j = 1.2$	8.178	3500 STEEL	495.30	0.202	4.76

The resulting safety factors are higher than 3, therefore the thickness and inner diameter are correct.

Table A10.13 Summary of the pipes that comply with the penstock conditions, based on the 175 m maximum head and the 0.080 m^3/s flow

Section	Material	Type of union	Class	d_e mm	t mm	d_i mm	l_i m	Quantity	Accessories
I	PVC	Rigid	7.5	219	7.9	203	60	13	1 Te of 219/110 mm PVC-C7.5 + 1 PVC-C7.5 pipe of 110 mm × 5 m
II	PVC	Rigid	10	219	10.4	198	60	13	1 curve – 219 mm PVC-C10×168°
III	PVC	Rigid	15	219	15.3	188	74	16	1 curve – 219 mm PVC-C15×165°
IV	Steel	Flange	Schedule 40	219	8.18	202	105.2	21	1 Steel curve Schedule 21 mm × 148°

Spreadsheet for penstock calculations

This consists of three parts:

The first part is the database entered into the spreadsheet for the corresponding process, including the flow design, gross head, total length of the pipe, etc. and the physical and mechanical characteristics of the selected piping material.

The second part consists of specific data on each section of the pipe; the information is processed in accordance with the studied formulas, to 'verify the resistance of the pipeline' to the static and dynamic pressures it is submitted to. The physical and mechanical characteristics of the selected material should meet safety standards.

All the information must be entered in the corresponding units so that the expected results are also expressed in those units.

The third part reveals the extent of the head loss and the power to be generated.

Below are copies of the spreadsheets for the two cases analysed and developed manually (the penstocks for the Sondor and Chetilla MHSs).

Table A10.14 Sizing of the PVC penstock, 'Sondor' MHS (Hb = 130.62 m)

A) Data

Design flow	m^3/s	0.15
Gross head	m	130.62
Length of pipe	m	251.75
Number of changes of direction		4
Number of changes of section		4
Fluid expansion module	kgf/cm^2	21,000
Kinematic viscosity of the water (15° C)	m^2	1.14E-06
Steel elasticity module	kgf/cm^2	2,000,000
PVC elasticity module	kgf/cm^2	28,000
Maximum traction effort of the steel	kgf/cm^2	3,500
Maximum traction effort of the PVC	kgf/cm^2	500
Steel surface roughness	mm	0.1
PVC surface roughness	mm	0.01

B) Evaluation of the resistance of the piping

		FU PVC (NTP ISO 4422)			
		Class 5	Class 7.5	Class 10	Class 15
Length of section	m	39.25	79	69.5	64
Height of the load in the section	m	28	56.75	87.5	130.6
Nominal diameter	inches	12	12	12	12
Outer diameter	m	0.315	0.315	0.315	0.315
Inner diameter	m	0.2996	0.2922	0.285	0.271
Thickness (consider corrosion, etc.)	mm	7.7	11.4	15	22
Water speed	m/s	2.13	2.24	2.35	2.60
Wave speed	m/s	258.47	315.76	363.62	443.78
Critical closing time	s	0.30	0.50	0.38	0.29
Maximum transitory pressure	m	56.06	72.00	87.16	117.64
Maximum pressure in the pipe	m	84.06	128.75	174.66	248.24
Thickness safety factor		3.06	3.03	3.01	3.27

C) Calculation of head loss

Relative roughness		3.34E-05	3.42E-05	3.51E-05	3.69E-05
Reynolds number		559,182.67	573,344.04	587,828.52	618,196.04
Friction factor		0.01338	0.01334	0.01331	0.01324
Head loss by friction	m	0.40	0.92	0.91	1.08
Secondary loss coefficient		1.20	1.20	1.20	1.20
Secondary head loss	m	0.28	0.31	0.34	0.41
Head loss in the section	m	0.68	1.23	1.25	1.49
Total head loss	m	3.16			
Percentage of losses	%	2.42			
Net head	m	127.46			
Mechanical power in the shaft	kW	140.52			
Electrical power in the alternator terminals	kW	120.15			

DESIGN OF THE RU PVC AND STEEL PIPE

Table A10.15 Sizing of the PVC and steel penstock, Chetilla MHS (Hb = 175 m)

A) Data

Design flow	m^3/s	0.08
Gross head	m	175
Total length of pipe	m	299.2
Number of changes of direction		4
Number of changes of section		4
Fluid expansion module	kgf/cm^2	21,000
Kinematic viscosity of the water (15° C)	m^2/s	1.14E-06
Steel elasticity module	kgf/cm^2	2,000,000
PVC elasticity module	kgf/cm^2	28,000
Maximum traction effort of the steel	kgf/cm^2	3,500
Maximum traction effort of the PVC	kgf/cm^2	500
Steel surface roughness	mm	0.1
PVC surface roughness	mm	0.01

B) Evaluation of the resistance of the piping

		RU PVC (national standards)			Steel
		Class 7.5	Class 10	Class 15	Schedule 40
Length of section	m	60	60	74	105.2
Height of the load in the section	M	34.3	69	121.5	175
Nominal diameter	inches	8	8	8	8
Outer diameter	M	0.219	0.219	0.219	0.219
Inner diameter	M	0.2032	0.1982	0.1884	0.202644
Thickness (consider corrosion, etc.)	mm	7.9	10.4	15.3	8.178
Water speed	m/s	2.47	2.59	2.87	2.48
Wave speed	m/s	315.24	363.11	443.85	1264.95
Critical closing time	s	0.38	0.33	0.33	0.17
Maximum transitory pressure	m	79.27	95.98	129.84	319.84
Maximum pressure in the pipe	m	113.57	164.98	251.34	494.84
Thickness safety factor		3.42	3.18	3.23	5.71

C) Calculation of head loss

Relative roughness		4.92E-05	5.05E-05	5.31E-04	4.93E-04
Reynolds number		439,714.24	450,807.99	474,257.66	440,921.73
Friction factor		0.01408	0.01405	0.01801	0.01785
Head loss by friction	m	1.29	1.46	2.97	2.91
Secondary loss coefficient		1.20	1.20	1.2	1.2
Secondary head loss	m	0.37	0.41	0.50	0.38
Head loss in the section	m	1.66	1.87	3.47	3.28
Total head loss	m	7.00			
Percentage of losses	%	4.00			
Net head	m	168.00			
Mechanical power in the shaft	kW	92.20			
Electrical power in the alternator terminals	kW	78.83			

EXAMPLE 11
Calculation of props or supports (see Chapter 8)

In the Chetilla MHS, which has a steel pipe section in the lower part, the supports were designed using the following field data:
Commercial steel pipe
$D_e = 0.219$ m
$D_i = 0.202$ m
Length of each pipe = 5 m
Pipe union
Distance between props = 5 m
$\alpha = 32°$
$\mu = 0.5$ friction coefficient between the steel and the concrete
$\mu_T = 0.45$ to 0.50, friction coefficient between the foundation soil and the concrete
$\sigma_T = 2$ kg-f/cm² (cohesive soil, hard clay)
$Q = 0.080$ m³/s

Preliminary sizing, in perspective

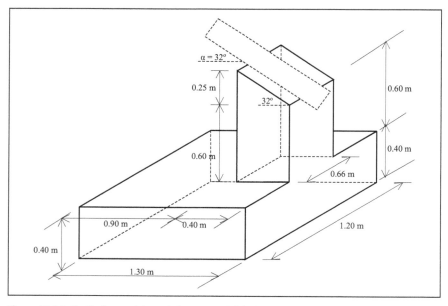

Figure A11.1 Preliminary sizing of the support block

CALCULATION OF PROPS OR SUPPORTS

Dimensions of the profile, for the corresponding calculations

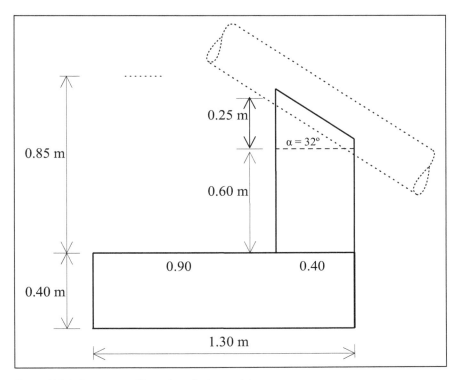

Figure A11.2 Support profile and preliminary sizing

Calculation of the weight of the pipe and the water per linear metre

$W = W_t + W_a$

a) W_t = weight of the pipe, per linear metre, in kg-f/m

$$W_t = 7860 \times \frac{3.1416}{4} [0.219^2 - 0.202^2] \times 1 = 44.18 \text{ kg-f/m}$$

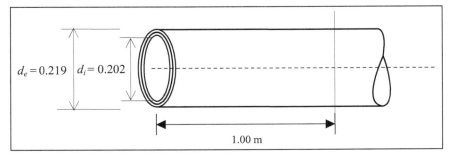

Figure A11.3 View of a section of penstock

b) W_a = weight of the water inside the pipe, per linear metre

$$W_a = 1000 \times \frac{3.1416 \times 0.202^2}{4} \times 1 \text{ m} = 32.04 \text{ kg-f/m}$$

Consequently, weight W of the pipe and the water per linear metre is:

$W = 44.18 + 32.04 = 76.22$ kg-f/m

The weight of the pipe and the water to be supported by the prop is:

$W_{ta} = W \times L_t = 76.22$ kg-f/m $\times 5$ m $= 381.10$ kg-f

With this value, components W_x, W_y and force of friction f were identified.

c) $W_x = W_{ta} \cdot \sin \alpha$

$W_x = 381.10 \times \sin 32° = 201.95$ kg-f

d) $W_y = W_{ta} \times \cos \alpha = 381.10 \times \cos 32° = 323.2$ kg-f
e) $f = \mu \cdot W_y \to 0.5 \times 323.2 = 161.6$ kg-f
f) Here, force f follows two directions. First, upwards in the direction of the pipe when the temperature makes the pipe expand, and, second, in the opposite direction when the pipe contracts. Consequently, the calculation is for the first case, when the pipe expands.

Case I: When the pipe expands, force of friction f moves upwards in the direction of the pipe (Figure A11.4)

To continue with the calculation, verify that the pipes joined to the flange acting as a prop, have no more than the admissible bending stress.

$$\delta_{admissible} = \frac{5 \text{ m}}{360} = 0.0138 \text{ m} = 13.8 \text{ mm}$$

Calculation of the maximum bending stress δ:

$$\delta = \frac{5 \times 76.22 \times 5^4}{384 \times 20 \times 10^9 \times I}$$

I = Moment of inertia of the section, in m^4

$$I = \frac{3.1416}{64} (0.219^4 - 0.202^4) = 3.51 \times 10^{-4} \text{ m}^4$$

Replacing the value of I, in δ

$$\delta = \frac{5 \times 76.22 \times 5^4}{384 \times 20 \times 10^9 \times 3.51 \times 10^{-4}} = 8.86 \times 10^{-5} \text{ m} = 0.0000886 \text{ m} = 0.08 \text{ mm}$$

Comparing 0.08 mm < 13.8 mm ... OK

CALCULATION OF PROPS OR SUPPORTS

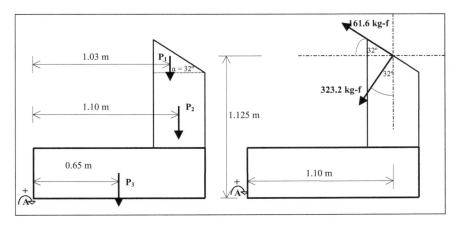

Figure A11.4 Breakdown of the weights of the concrete block and of acting forces

Verifying the stability of the support

Bending

Condition, $\dfrac{\Sigma StabilizingM}{\Sigma BendingM} \geq 2$

a) To find the Σ of the stabilizing moments, calculate the partial weights of the block (Figures A11.2 and A11.4)

P_i = specific weight of the concrete × volume

Table A11.1 Weights for the different blocks

N°	Areas (m²)	Volume (m³)	Weights (kg-f)
1	(0.25 × 0.40)/2 = 0.05	0.033	72.6
2	0.40 × 0.60 = 0.24	0.1584	348.48
3	0.40 × 1.30 = 0.52	0.624	1372.80
Total	0.81	0.8154	W_c = 1793.88

$\Sigma StabilizingM$ in A = P_1 × 1.03 + P_2 × 1.10 + P_3 × 0.65 + 323.2 × cos 32° × 1.10 − 161.6 × sin 32° × 1.10
= 72.6 × 1.03 + 348.48 × 1.10 + 1372.8 × 0.65 + 274.08 × 1.10 − 85.63 × 1.10
= 1557.72 kg-f/m

$\Sigma BendingM$ in A = − 161.6 × cos 32° × 1.125 − 323.2 × sin 32° × 1.125 =
−154.17 −192.67
= − 346.85 kg-f/m

$\dfrac{1557.72}{346.85} \geq 2$

4.49 > 2 ... OK

Sliding

Condition: $\Sigma F_x < \mu_T \cdot \Sigma F_y$
Calculation of $\Sigma F_x = -161.6 \times \cos 32° - 323.2 \times \sin 32° = -308.3$ kg-f
Calculation of $\Sigma F_y = -P_1 - P_2 - P_3 + 161.6 \times \sin 32° - 323.2 \times \cos 32°$
$= -72.6 - 348.48 - 1372.8 + 85.63 - 274.08$
$= -1982.33$ kg-f

Applying the condition: $\Sigma F_x < \mu_T \Sigma F_y$; $(\mu_T = 0.5)$

308.3 kg-f $< 0.5 \times 1982.33$
308.3 kg-f < 991.16 kg-f ... OK

Stability of the foundation soil

Calculation of X_g = moment length for the centre of gravity of the concrete block, applying the area method:

From Figure A11.5: $X_G = \dfrac{\Sigma areaM}{\Sigma areas}$

$$X_G = \frac{0.05 \times 1.03 + 0.24 \times 1.10 + 0.52 \times 0.65}{0.05 + 0.24 + 0.52} = 0.806 \text{ m}$$

Calculating R_y and X

In Figure A11.6, apply the sum of the moments in point A, with the intervention of the resulting forces acting in the support:

$\Sigma M_A: -R_y \cdot X - \Sigma F_x \cdot 1.125 + F_y \cdot 1.10 + W_c \cdot X_G = 0$

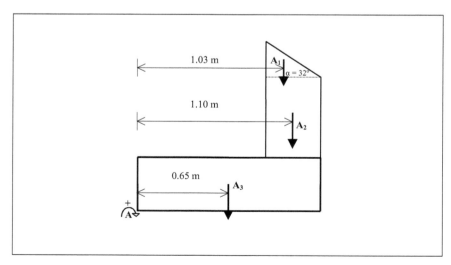

Figure A11.5 Areas and lever arms to determine x_G of the block

CALCULATION OF PROPS OR SUPPORTS   321

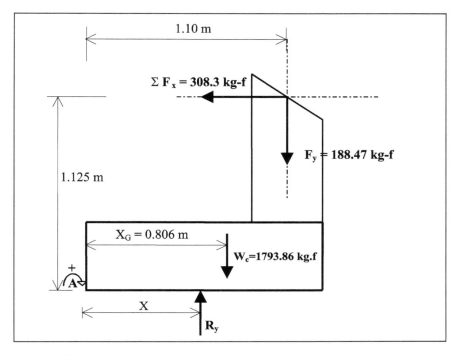

Figure A11.6 Resulting forces for calculating the stability of the support

Calculation of F_y, $\rightarrow F_y = \Sigma F_y = W_c = 1982.33 - 1793.86 = 188.47$ kg-f

Replacing values:
$-R_y.X - 308.3 \times 1.125 + 188.47 \times 1.10 + 1793.86 \times 0.806 = 0$
$R_y.X = -346.84 + 207.32 + 1445.85$
$R_y.X = 1306.84$ kg-f.m

Calculation of R_y:
$R_y = W_c + F_y = 1793.86 + 188.47$
$R_y = 1982.33$ kg-f

Therefore: $X = 1306.84/1982.33$
$X = 0.659$ m

Eccentricity: $e = X - b/2$
$e = 0.659 - 1.30/2 = 0.659 - 0.65 = 0.009$ m

Calculation of σ_{max}:
$$\sigma_{max} = \frac{R_y}{A}\left[1 + \frac{6e}{b}\right] = \frac{1982.33 \text{ kg-f}}{130 \times 120 \text{ cm}^2}\left[1 + \frac{6 \times 0.009}{1.30}\right] = 0.13 \text{ kg-f/cm}^2$$

Calculation of σ_{min}:
$$\sigma_{min} = \frac{R_y}{A}\left[1 - \frac{6e}{b}\right] = \frac{1982.33 \text{ kg-f}}{130 \times 120 \text{ cm}^2}\left[1 - \frac{6 \times 0.009}{1.30}\right] = 0.12 \text{ kg-f/cm}^2$$

Verification: $\sigma_T > \sigma_{max} > \sigma_{min}$
2 kg-f/cm² > 0.13 kg-f/cm² > 0.12 kg-f/cm² ... OK

The shape and pre-established dimensions of the support comply with all three stability conditions when the pipe expands. Now it must be proved that it is also stable when the pipe contracts.

CASE II: When the pipe contracts, force of friction f acts downwards in the direction of the pipe (Figure A11.7)

The values calculated in the first case remain the same; only the values that depend on the force of friction vary, due to the change in direction.
$W = 76.22$ kg-f/m
$W_{ta} = 76.22$ kg-f/m × 5m = 381.10 kg-f
$W_x = 201.95$ kg-f (has no part in the calculation of the support)
$W_y = 323.2$ kg-f
$f = \mu.W_y \rightarrow 0.5 \times 323.19 = 161.6$ kg-f (downwards in the direction of the pipe)
$I = 3.51 \times 10^{-4}$ m⁴
$\delta = 0.08$ mm
$\delta_{admissible} = 13.8$ mm
$\delta < \delta_{admissible}$

Verifying the stability of the support

Bending

Condition, $\dfrac{\Sigma StabilizingM}{\Sigma BendingM} \geq 2$

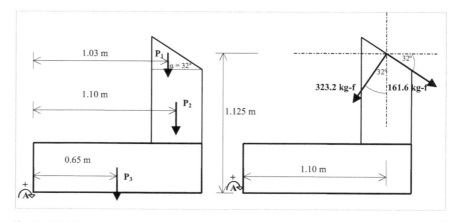

Figure A11.7 Breakdown of the weights of the concrete block and change of direction of the force of friction (Case II)

CALCULATION OF PROPS OR SUPPORTS

$\Sigma StabilizingM$ in $A = P_1 \times 1.03 + P_2 \times 1.10 + P_3 \times 0.65 + 323.2 \times \cos 32° \times 1.10 + 161.6 \times \sin 32° \times 1.10$
$= 72.6 \times 1.03 + 348.48 \times 1.10 + 1372.8 \times 0.65 + 274.08 \times 1.10 + 85.63 \times 1.10$
$= 1746.08$ kg-f.m
$\Sigma BendingM$ in $A = 161.6 \times \cos 32° \times 1.125 - 323.2 \times \sin 32° \times 1.125 = 154.17 - 192.67$
$= -38.50$ kg-f.m

$\dfrac{1746.08}{38.50} \geq 2$

$45.35 > 2 \ldots$ OK

Sliding

Condition: $\Sigma F_x < \mu_T \cdot \Sigma F_y$

Calculation of $\Sigma F_x = 161.6 \times \cos 32° - 323.2 \times \sin 32° = -34.3$ kg-f
Calculation of $\Sigma F_y = -P_1 - P_2 - P_3 - 161.6 \times \sin 32° - 323.2 \times \cos 32°$
$= -72.6 - 348.48 - 1372.8 - 85.63 - 274.08$
$= -2153.57$ kg-f

Applying the condition: $\Sigma F_x < \mu_T \cdot \Sigma F_y$; ($\mu_T = 0.5$, Table 2.2)

34.3 kg-f $< 0.5 \times 2153.57$
34.3 kg-f < 1076.78 kg-f $\ldots$ OK

Stability of the foundation soil

$X_g = 0.806$ m

Calculating R_y and X
In Figure A11.8, apply the sum of the moments in point A, with the intervention of the resulting forces acting in the support.
$\Sigma M_A: -R_y \cdot X - F_x \times 1.125 + F_y \times 1.10 + W_c \cdot X_G = 0$

Calculation of F_y, $\rightarrow F_y = \Sigma F_y - W_c = 2153.57 - 1793.86 = 359.71$ kg-f

Replacing values:
$-R_y \cdot X - 34.3 \times 1.125 + 359.71 \times 1.10 + 1793.86 \times 0.806 = 0$
$R_y \cdot X = -38.58 + 395.68 + 1445.85$
$R_y \cdot X = 1802.95$ kg-f.m

Calculation of R_y:
$R_y = W_c + F_y = 1793.86 + 359.71$
$R_y = 2153.57$ kg-f

Therefore: $X = 1802.95/2153.57$
$X = 0.837$ m

Eccentricity: $e = X - b/2$
$e = 0.837 - 1.30/2 = 0.837 - 0.65 = 0.187$ m

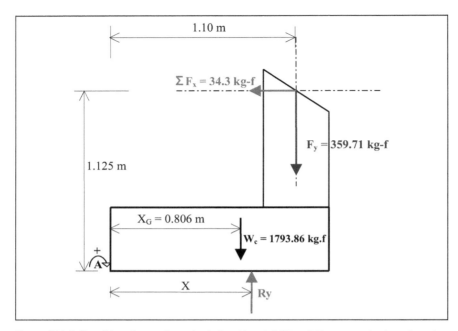

Figure A11.8 Resulting forces for calculating the stability of the support when the pipe contracts

Calculation of σ_{max}:

$$\sigma_{max} = \frac{R_y}{A}\left[1 + \frac{6e}{b}\right] = \frac{2153.57 \text{ kg-f}}{130 \times 120 \text{ cm}^2}\left[1 + \frac{6 \times 0.187}{1.30}\right] = 0.26 \text{ kg-f/cm}^2$$

Calculation of σ_{min}:

$$\sigma_{min} = \frac{R_y}{A}\left[1 - \frac{6e}{b}\right] = \frac{2153.57 \text{ kg-f}}{130 \times 120 \text{ cm}^2}\left[1 - \frac{6 \times 0.187}{1.30}\right] = 0.18 \text{ kg-f/cm}^2$$

Verification: $\sigma_T > \sigma_{max} > \sigma_{min}$
2 kg-f/cm^2 > 0.26 kg-f/cm^2 > 0.18 kg-f/cm^2 ... OK

It was thus proved that the dimensions and shape of the concrete block comply with the stability conditions for both cases: when the pipe expands and when it contracts. Therefore, it was accepted as a typical support for the steel section of the pipe.

Calculations procedure

This consists of two parts:
   The first is the database entered in accordance with the design and preliminary sizing, established for verification purposes.

CALCULATION OF PROPS OR SUPPORTS

The information entered is used to calculate the forces intervening on the support, given the weight of the pipe and the water. Likewise, the weight of the concrete block itself is calculated.

The second part evaluates the stability conditions of the support, its resistance to sliding and the stability of the soil for two cases: when the pipe expands and when it contracts.

If the project designer changes the profile, the compatibility of its application must be reviewed or modified.

Below it can be seen a diagram showing the dimensions of the support in generalized manner and a table with the data for the Chetilla MHS.

Table A11.2 Design of the support for the steel pipe: Chetilla MHS

Description	Value	
H	175.000	H = static head pressure (m)
Q	0.080	Q = design flow (m^3/s)
D_i	0.202	D_i = inner diameter of the pipe (m)
T	0.009	t = thickness of the pipe (m)
D_e	0.219	D_e = outer diameter of the pipe (m)
A	32.000	α = angle formed by the pipe with the horizontal
ρ_t	7860.000	ρ_t = specific weight of the pipe (kg-f/m^3)
ρ_a	1000.000	ρ_a = specific weight of the water (kg-f/m^3)
W_t	44.182	W_t = weight of the pipe (kg-f/m)
W_a	32.047	W_a = weight of the water (kg-f/m)
L_t	5.00	L_t = length of the pipe (m)
u	0.5	u = friction coefficient between the steel and C°
S_{adm}	2.00	S_{adm} = admissible bearing capacity of the soil kg-f/cm^2
Data		
Wy	323.23	
Wx	201.98	
f	161.62	
a	0.90	
b	1.30	
c	0.40	
d	0.60	
e	0.25	
f	0.40	
g	0.66	
h	1.20	

EXAMPLE 12
Calculations for the anchor (see Chapter 8)

In the Chetilla MHS, because the section of the pipe in the lower section is steel, the following anchor was calculated for both cases, i.e. for when the pipe expands or when it contracts. The field data, the characteristics of the materials, etc. appear in Figure A12.1.

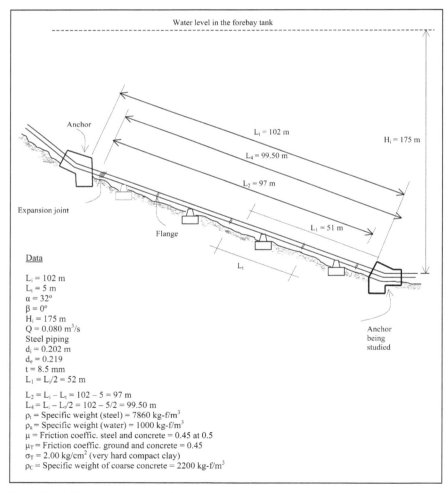

Data

$L_i = 102$ m
$L_t = 5$ m
$\alpha = 32°$
$\beta = 0°$
$H_i = 175$ m
$Q = 0.080$ m^3/s
Steel piping
$d_i = 0.202$ m
$d_e = 0.219$
$t = 8.5$ mm
$L_1 = L_t/2 = 52$ m
$L_2 = L_i - L_t = 102 - 5 = 97$ m
$L_4 = L_i - L_t/2 = 102 - 5/2 = 99.50$ m
ρ_t = Specific weight (steel) = 7860 kg-f/m^3
ρ_a = Specific weight (water) = 1000 kg-f/m^3
μ = Friction coeffic. steel and concrete = 0.45 at 0.5
μ_T = Friction coeffic. ground and concrete = 0.45
$\sigma_T = 2.00$ kg/cm^2 (very hard compact clay)
ρ_C = Specific weight of coarse concrete = 2200 kg-f/m^3

Figure A12.1 Field data and data from steel pipe catalogues to calculate the anchor for the lower part

Case I: When the pipe expands

Preliminary sizing of the profile to prove the stability conditions

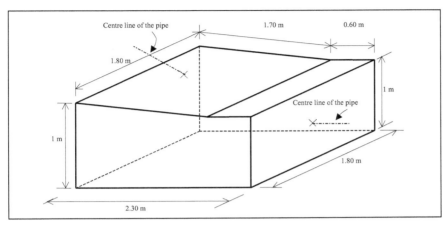

Figure A12.2 Preliminary sizing of the concrete anchor

Calculation of forces

Since this is the same section of steel piping used for calculating the supports, the following can be used:
Weight of the pipe per linear metre: $W_t = 44.10$ kg-f/m,
Weight of the water per linear metre: $W_a = 32.04$ kg-f/m,
Weight of the pipe and the water per metre $W = 76.18$ kg-f/m, and other characteristics noted in Figure A12.1, such as the inner diameter, thickness, flow and pressure head.
F_1 = pipe weight component + water weight

$F_1 = W.L_1.\cos \alpha$
$L_1 = L_i/2 \rightarrow L_1 = 102/2 = 51$ m
$F_1 = 76.22 \times 51 \times \cos 32° = 3{,}296.5$ kg-f

F_2 = force of friction between the pipe and the supports

$F_2 = \mu.W.L_2.\cos \alpha \rightarrow 0.5 \times 76.22 \times 97 \times \cos 32° = 3135$ kg-f

F_3 = hydrostatic pressure in the elbow or in the curve, when changing direction

$$F_3 = 1.6 \times 10^3 \times H_i \times d_i^2 \times \sin \frac{\beta - \alpha}{2}$$

$F_3 = 1.6 \times 10^3 \times 175 \times 0.202^2 \times \sin \dfrac{0 - 32°}{2} = -3{,}149.2$ kg-f

F_4 = pipe weight component, parallel to the pipe
$F_4 = W_t L_4 \times \sin \alpha \rightarrow = 44.18 \times 99.50 \times \sin 32° = 2329.50$ kg-f
F_5 = force caused by temperature changes in the pipeline

$F_5 = 31 \times d_i \times t \times E \times a' x \Delta T$

Because this section of the pipe has an expansion joint:

F_6 = force of friction in the expansion joint

$F_6 = 10 \times d_i \rightarrow = 10 \times 202 = 2020$ kg-f

F_7 = force due to the hydrostatic pressure within the expansion joint

$F_7 = 3.1 \times H_i \times d_i \times t \rightarrow = 3.1 \times 175 \times 0.202 \times 8.5 = 931.5$ kg-f

F_8 = force caused by a change of direction of the quantity of water moving in the elbow or curve. The direction is the same as F_3

$$F_8 = 250 \times \left(\frac{Q}{d_i}\right)^2 \times \sin\frac{\beta - \alpha}{2} = 250 \times \left(\frac{0.080Q}{0.202}\right)^2 \times \sin\frac{0 - 32}{2} = -10.80 \text{ kg-f}$$

F_9 = force caused by a change in the inner diameter, with a reducer,

$$F_9 = 0$$

because there is no reduction

Forces acting on the anchor (Figure A12.3)

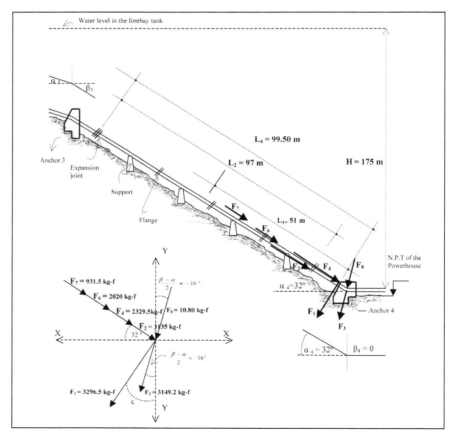

Figure A12.3 Acting forces for calculating Anchor 4, when the pipe expands

Verifying the stability of the anchor (Figure A12.4)

Calculating the weight W_c of the anchor

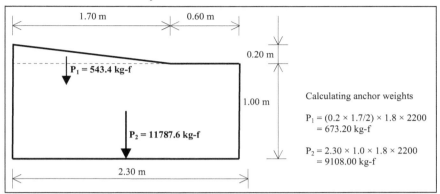

Figure A12.4 Calculating the weight of the concrete anchor

Representation of acting forces to verify the stability of the anchor (Figure A12.5)

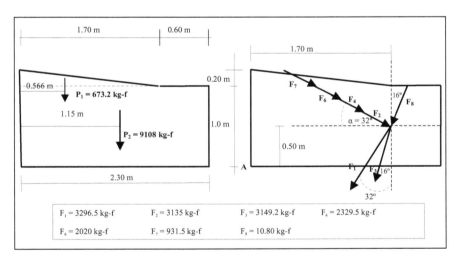

Figure A12.5 Diagram of concrete weights and acting forces

Bending stability

Condition: $\dfrac{\Sigma StabilizingM}{\Sigma BendingM} \geq 2$

$\Sigma\text{stab.}M_A = P_1 \times 0.566 + P_2 \times 1.15 + F_1 \times \cos 32° \times 1.70 + (F_3 + F_8)\cos 16° \times 1.70 + (F_2 + F_4 + F_6 + F_7)\sin 32° \times 0.50$

= 673.2 × 0.566 + 9108 × 1.15 + 3296.5 × cos32° × 1.70 + (3149.2 + 10.80) × cos16° × 1.70 + (3135 + 2329.5 + 2020 + 931.5) × sin 32° × 0.5 = 28353.29 kg-f.m

$\Sigma ben.M_A = - F_1 \times sin32° \times 0.50 - (F_3 + F_8) sin16° \times 0.50 + (F_2 + F_4 + F_6 + F_7) \times cos32° \times 0.50$
= −3296.5sin32° × 0.50 − (3149.2 + 10.80)sin16° × 0.50 + (3135 + 2329.5 + 2020 + 931.5) × cos32° × 0.50
= −873.44 − 435.50 + 3568.58 = 2259.64 kg-f.m

Verification: $\dfrac{28353.29}{2259.64} \geq 2$

12.54 > 2 ... OK

Verification of sliding stability

Condition: $\Sigma F_x < \mu_T \cdot \Sigma F_y$

$\Sigma F_x = - F_1 \times sin32° - (F_3 + F_8)sin16° + (F_2 + F_4 + F_6 + F_7) \times cos32°$
= −3296.5sin32° − (3149.2 + 10.80)sin16° + (3135 + 2329.5 + 2020 + 931.5) × cos32°
= −1746.9 − 871.01 + 7137.17 = 4519.3 kg-f
$\Sigma F_y = - F_1 \times cos32° - (F_3 + F_8)cos16° - (F_2 + F_4 + F_6 + F_7)sin32° - W_c$
= −3296.5 × cos32° − (3149.2 + 10.80) × cos16° − (8416) × sin32° − 9781.2
= −2795.6 − 3037.6 − 4459.8 − 9781.2 = −20074.2 kg-f.

$\mu_T = 0.5$
4519.3 kg-f < 0.5 × 20074.2
4519.3 kg-f < 10037.1 kg-f ... OK

Stability of the foundation soil

Calculation of X_G = Abscissa of the centre of gravity in the concrete block

$X_G = \dfrac{\Sigma M\ areas}{\Sigma areas}$ $X_G = \dfrac{0.17 \times 0.566 + 2.30 \times 1.15}{2.47} = 1.109$ m

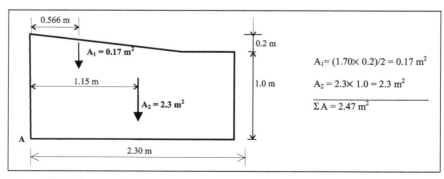

Figure A12.6 Diagram of areas

CALCULATION FOR THE ANCHOR

Calculation of X and R_y
From Figure A12.7:
$\Sigma M_A: -R_y \cdot X + W_c \times X_G + F_x \times 0.50 + F_y \times 1.70 = 0$
$R_y \cdot X = 9781.2 \times 1.109 + 4519.3 \times 0.50 + 10293 \times 1.70$
$R_y \cdot X = 30605.10$ kg-f.m

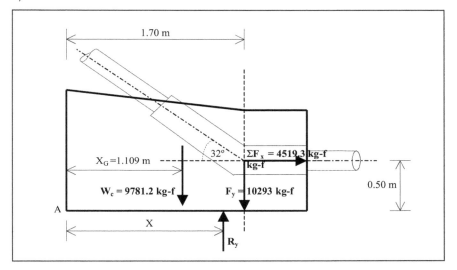

Figure A12.7 Diagram of resulting forces for calculating the soil stability

From Figure A12.7: $R_y = W_c + F_y$
$R_y = 9781.2 + 10293$
$R_y = 20074.2$ kg-f
Then $X = 30605.1/20074.2$
$X = 1.524$ m
Therefore, eccentricity $e = X - b/2$
$e = 1.524 - 2.30/2$
$e = 0.374$ m

Calculation of σ_{max}:
$\sigma_{max} = \dfrac{R_y}{A}\left[1 + \dfrac{6e}{b}\right] = \dfrac{20074.2 \text{ kg-f}}{230 \times 180 \text{ cm}^2}\left[1 + \dfrac{6 \times 0.374}{2.30}\right] = 0.96$ kg-f/cm^2

Calculation of σ_{min}:
$\sigma_{min} = \dfrac{R_y}{A}\left[1 - \dfrac{6e}{b}\right] = \dfrac{20074.2 \text{ kg-f}}{230 \times 180 \text{ cm}^2}\left[1 - \dfrac{6 \times 0.374}{2.30}\right] = 0.01$ kg-f/cm^2

Verification: $\sigma_T > \sigma_{max} > \sigma_{min}$
2 kg-f/cm^2 > 0.96 kg-f/cm^2 > 0.01 kg-f/cm^2
... OK

So far, it has been proved that the anchor is stable, given its shape and assumed dimensions when the pipe expands.

The next step is to prove that the same anchor is stable when the pipe contracts.

Case II: When the pipe contracts

In this case, only forces F_2 and F_6 change direction; the magnitudes are the same and the diagram of acting forces is as shown in Figure A12.8.

Continue with the same calculation procedure:

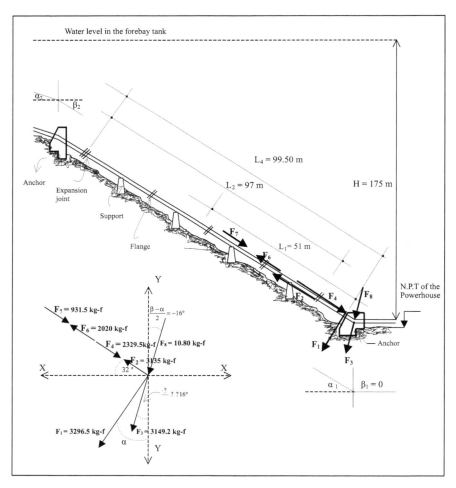

Figure A12.8 Acting forces for calculating the anchor when the pipe contracts

CALCULATION FOR THE ANCHOR

Verifying the stability of the anchor (Figure A12.9)

Bending stability

Condition: $\dfrac{\Sigma StabilizingM}{\Sigma BendingM} \geq 2$

$\Sigma \text{stab.}M_A = P_1 \times 0.566 + P_2 \times 1.15 + F_1 \times \cos 32° \times 1.70 + (F_3 + F_8)\cos 16° \times 1.70 + (-F_2 + F_4 - F_6 + F_7)\sin 32° \times 1.70$
$= 673.2 \times 0.566 + 9108 \times 1.15 + 3296.5 \times \cos 32° \times 1.70 + (3149.2 + 10.80) \times \cos 16° \times 1.70 + (-3135 + 2329.5 - 2020 + 931.5) \times \sin 32° \times 1.70$
$= 381.03 + 10474.02 + 4752.50 + 5163.89 - 1706.23 = 19065.40$ kg-f/m

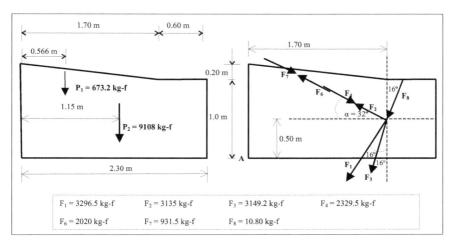

Figure A12.9 Concrete weights and acting forces when the pipe contracts

$\Sigma \text{ben}M_A = - F_1 \times \sin 32° \times 0.50 - (F_3 + F_8)\sin 16° \times 0.50 + (-F_2 + F_4 - F_6 + F_7) \times \cos 32° \times 0.50$
$= -3296.5 \sin 32° \times 0.50 - (3149.2 + 10.80)\sin 16° \times 0.50 + (-3135 + 2329.5 - 2020 + 931.5) \times \cos 32° \times 0.50$
$= -873.43 - 435.50 - 803.10 = -2112.03$ kg-f/m

Verification: $\dfrac{19065.04}{2112.03} \geq 2$

$9.02 > 2 \ldots$ OK

Verification of sliding stability

Condition: $\Sigma F_x < \mu_T \cdot \Sigma F_y$

$\Sigma F_x = - F_1 \times \sin 32° - (F_3 + F_8)\sin 16° + (-F_2 + F_4 - F_6 + F_7) \times \cos 32°$
$= -3296.5 \sin 32° - (3149.2 + 10.80)\sin 16° + (-3135 + 2329.5 - 2020 + 931.5) \times \cos 32°$

$= -1746.9 - 871.01 - 1606.2 = -4224.11$ kg-f
$\Sigma F_y = -F_1 \times \cos 32° - (F_3 + F_8)\cos 16° + (F_2 - F_4 + F_6 - F_7)\sin 32° - W_c$
$= -3296.5 \times \cos 32° - (3149.2 + 10.80) \times \cos 16° + (3135 - 2329.5 + 2020 - 931.5)$
$\times \sin 32° -9781.2$
$= -2795.6 - 3037.6 + 1003.66 - 9781.2 = -14610.74$ kg-f

$\mu_T = 0.5$ for hard clay (Table 2.2)
4224.11 kg-f < 0.5 × 14610.74 kg-f
4224.11 kg-f < 7305.37 kg-f ... OK

Stability of the foundation soil (Figure A12.10)

The concrete block is the same, $X_G = 1.109$ m
SF_x changes in value and direction
F_y changes only in value

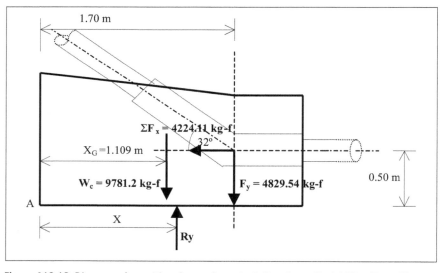

Figure A12.10 Diagram of resulting forces for calculating the soil stability (Case II)

Calculation of X and R_y
$\Sigma M_A = -R_y.X + W_c \times X_G - F_x \times 0.50 + F_y \times 1.70 = 0$
$= -R_y.X + 9781.2 \times 1.109 - 4224.11 \times 0.50 + 4829.54 \times 1.70 = 0$
$R_y.X = 10847.35 - 2112.05 + 8210.21$
$R_y.X = 16945.51$ kg-f.m
From Figure A12.10: $R_y = W_c + F_y$
$R_y = 9781.2 + 4829.54$
$R_y = 14610.74$ kg-f
Then $X = 16945.51 / 14510.74$
$X = 1.159$ m

Eccentricity $e = X - b/2$
$e = 1.159 - 2.30/2$
$e = 0.009$ m

Calculation of σ_{max}:

$$\sigma_{max} = \frac{R_y}{A}\left[1 + \frac{6e}{b}\right] = \frac{14610.74 \text{ kg-f}}{230 \times 180 \text{ cm}^2}\left[1 + \frac{6 \times 0.009}{2.30}\right] = 0.36 \text{ kg-f/cm}^2$$

Calculation of σ_{min}:

$$\sigma_{min} = \frac{R_y}{A}\left[1 - \frac{6e}{b}\right] = \frac{14610.74 \text{ kg-f}}{230 \times 180 \text{ cm}^2}\left[1 - \frac{6 \times 0.009}{2.30}\right] = 0.34 \text{ kg-f/cm}^2$$

Verification: $\sigma_T > \sigma_{max} > \sigma_{min}$
2 kg-f/cm^2 > 0.36 kg-f/cm^2 > 0.34 kg-f/cm^2 ... OK

Having made the calculations and proved that the anchor complies with the stability conditions in both cases, the established dimensions and shape were accepted.

In this case, the project designer could try out other shapes and profiles for the supports and anchors in an effort to reduce costs, but without posing a risk to the safety of the works.

After observing the profile of the pipeline for the Chetilla MHS project, an inward facing anchor was calculated. The same procedure is applicable for the calculation of an outward facing anchor, verifying that the profile and preliminary sizing established for the selected block complies with the three stability conditions: bending, sliding and foundation soil.

In turn, these stability conditions are subject to the acting forces, depending on whether the pipeline is steel or PVC.

Example of the calculations for an outward facing anchor

In Figure A12.11, anchor 2 corresponds to this type.
The data required was obtained from
the design
$Q = 0.080$ m^3/s
and from the calculations made to determine the diameter and type of pipe
PVC, rigid union, Class 10, Peruvian Technical Standard ITINTEC
$d_e = 219$ mm
$t = 10.4$ mm
$d_i = 198.2$ mm
$\gamma_t = 1450$ kg-f/m^3
$L_t = 5$ m stop pin, bell
$\mu = 0.4$ friction coefficient between the PVC and the soil
Foundation soil (compact clay at the site of anchor 2)
$\sigma_T = 1$ kg-f/cm^2

From the pipeline profile:
$H_i = 69$ m
$L_1 = 30$ m
$L_2 = 55$ m
$L_4 = 57.50$
$\alpha = 36°$
$\beta = 46°$

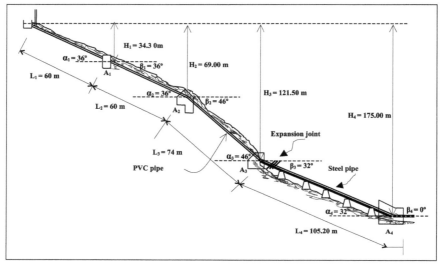

Figure A12.11 Profile of the penstock with steel pipes in the lower part and PVC pipes in the higher part

Shape and size selection

Select the shape and preliminary sizing of the anchor, bearing in mind the change of direction of the pipeline in accordance with angle $\alpha°$ and $\beta°$, as shown in Figure A12.12 (a) and (b); then verify the stability when the pipe expands or contracts, depending on the acting forces in the PVC pipes.

Calculation of W = W$_t$ + W$_a$

a) $W_t = \gamma_t \times \dfrac{\pi}{4}\left[d_e^2 - d_i^2\right] = 1450 \times \dfrac{3.1416}{4} [0.219^2 - 0.1982^2] = 9.88$ kg-f/m

b) $W_a = \gamma_t \times \dfrac{\pi}{4} \times d_i^2 = 1000 \times \dfrac{3.1416}{4} \times 0.1982^2 = 30.85$ kg-f/m

Then $W = 9.88 + 30.85 = 40.73$ kg-f/m

CALCULATION FOR THE ANCHOR

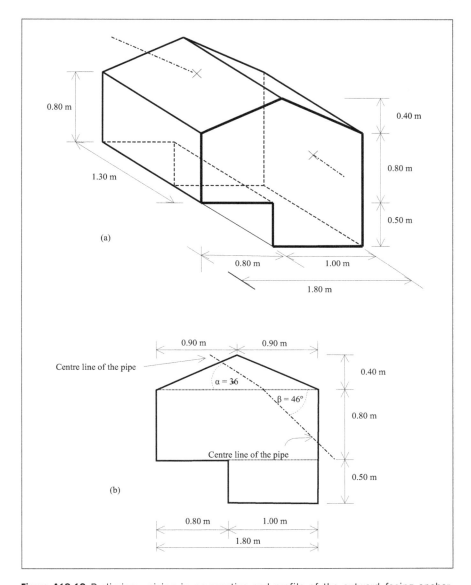

Figure A12.12 Preliminary sizing in perspective and profile of the outward facing anchor

Calculation of acting forces, in kg-f (Figure A12.13)

As in this case the pipeline consists of buried RU PVC pipes, by definition, forces F5, F6, F7 and F9 = 0. Therefore, the acting forces are:
a) $F_1 = W. L_1.\cos 36° = 40.73 \times 30 \times \cos 36° = 988.53$
b) $F_2 = \mu.W.L_2.\cos 36° = 0.4 \times 40.73 \times 55.00 \times \cos 36° = 724.92$

c) $F_3 = 1.6 \times 10^3 \times H_i \times d_i^2 \times \sin(46° - 36°)/2 = 1600 \times 69 \times 0.1982^2 \times \sin 5° = 377.98$
d) $F_4 = W_t \times L_4 \times \sin 36° = 9.88 \times 57.5 \times \sin 36° = 334.00$
e) $F_8 = 250(Q/d_i)^2 \times \sin(46° - 36°)/2 = 250(0.08/0.1982)^2 \times \sin 5° = 3.55$

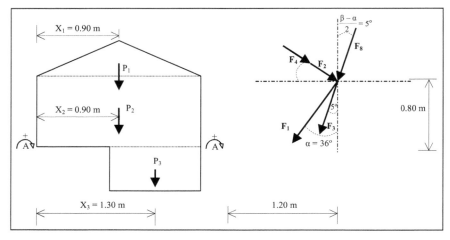

Figure A12.13 Diagram of partial weights of the concrete anchor and acting forces (Case I, when the pipe expands)

The stability was verified taking the following into consideration:
a) The elbow or the point where the pipe changes direction and the forces act, is located horizontally at $(2/3)b$ and vertically at distance d from point A.
Distances b and d are the pre-established dimensions for the anchor.
b) The stabilizing moments (vertical) and the bending moments were calculated with respect to point A.
c) The auxiliary calculations to verify the stability of the concrete block are summarized in Table A12.1.

Table A12.1 Summary of calculations of the stability of concrete blocks

N°	Area (m²)	Volume (m³)	Weights: P_i (kg-f)	X_i (m)	$A_i \cdot X_i M^2$ (m)	$P_i \cdot X_i$ (kg-f.m)
1	0.36	0.468	1029.6	0.90	0.324	926.64
2	1.44	1.872	4118.4	0.90	1.296	3706.56
3	0.50	0.65	1430.0	1.30	0.65	1859.0
Total	2.30	2.99	6578.0			6492.2

CALCULATION FOR THE ANCHOR

Verifying the stability of the blocks

Bending

$\Sigma \text{stabilizingM} / \Sigma \text{bendingM} > 2$

$\Sigma \text{stabilizingM} = P_1.X_1 + P2.X_2 + P_3 \times X_3 + F_1.\cos 36° \times 1.20 + (F_3 + F_8)\cos 5° \times 1.20$
$+ (F_2 + F_4)\sin 36° \times 1.20$
$= 6492.2 + 988.7 \times \cos 36 \times 1.20 + 381.53 \times \cos 5° \times 1.20 + 1059.02 \times \sin 36° \times 1.20$
$= 6492.20 + 959.85 + 456.09 + 746.97 = 8655.11 \text{ kg-f.m}$

$\Sigma \text{bendingM} = -F_1 \times \sin 36° \times 0.80 - (F_3 + F_8)\sin 5° \times 0.80 + (F_2 + F_4)\cos 36° \times 0.80$
$= -988.7 \times \sin 36° \times 0.80 - 381.53 \times \sin 5° \times 0.80 + 1059.02 \times \cos 36° \times 0.80$
$= -464.91 - 26.60 + 685.41 = 193.90 \text{ kg-f/m}$
Then: $8655.11/193.90 > 2$
$44.63 > 2 \ldots \text{OK}$

Sliding

$\Sigma F_x < \mu.\Sigma F_y$
$\Sigma F_x = -F_1 \times \sin 36° - (F_3 + F_8)\sin 5° + (F_2 + F_4)\cos 36°$
$= -988.7 \times \sin 36° - 381.53 \times \sin 5° + 1059.02 \times \cos 36°$
$= -581.14 - 33.25 + 856.76 = 242.37 \text{ kg-f}$
$\Sigma F_y = -P_1 - P_2 - P_3 - F_1.\cos 36° - (F_3 + F_8)\cos 5° - (F_2 + F_4)\sin 36°$
$= -6578 - 988.7 \times \cos 36° - 381.53 \times \cos 5 - 1059.02 \times \sin 6°$
$= -6578 - 799.87 - 380.07 - 622.47 = -8380.41 \text{ kg-f}$
confirming: $242.37 \text{ kg-f} < 0.5 \times 8380.141 \text{ kg-f}$
$242.37 < 4190.20 \ldots \text{OK}$

Soil stability:

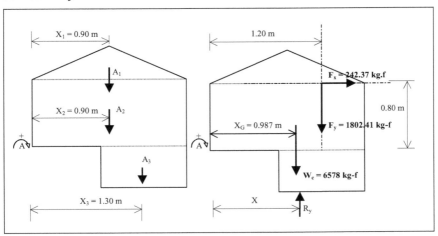

Figure A12.14 Diagram of partial weights of the concrete anchor and acting forces (Case I, when the pipe expands)

Calculation of X_G, for the concrete block, with the areas method (Figure A12.14)

$$X_G = \frac{A_1 \times X_1 + A_2 \times X_2 + A_3 \times X_3}{\Sigma A_i} = \frac{2.27}{2.30} = 0.987 \text{ m}$$

Calculation of R_y, and X (Figure A12.14)
Apply ΣM_A: $-R_y.X + 6578 \times 0.987 + 242.37 \times 0.80 + 1802.41 \times 1.20 = 0$
$R_y.X = 8849.27$ kg-f/m
In the figure: $R_y = W_c + F_y = 6578 + 1802.41 = 8380.41$ kg-f
$X = 8849.27/8380.41 = 1.055$ m

Calculating the eccentricity factor $e = X - b/2 \rightarrow 1.055 - 1.80/2 = 0.156$ m
Calculation of the maximum and minimum bearing capacity of the soil:

$$\sigma = \frac{R_y}{A}\left[1 \pm \frac{6 \times e}{b}\right] = \frac{8380.41}{180 \times 130}\left[1 \pm \frac{6 \times 0.156}{1.80}\right] = 0.400872 \, [1 \pm 0.52]$$

a) $\sigma_{max} = 0.400872(1.52) = 0.609 = 0.61$ kg-f/cm²
b) $\sigma_{min} = 0.400872(0.48) = 0.192 = 0.19$ kg-f/cm²

Confirming that: 1 kg-f/cm² > 0.61 kg-f/cm² > 0.19 kg-f/cm² ... OK

These calculations prove that the anchor is stable. The next step is to prove that it is also stable in the second case, i.e. when the pipe contracts.

Case II: When the pipe contracts

The intervening forces are the same. Only force F_2 changes direction (Figure A12.15)

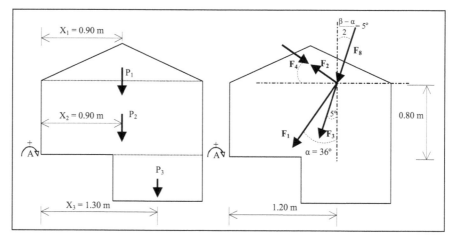

Figure A12.15 Diagram of partial weights of the concrete anchor and acting forces (Case II, when the pipe contracts)

CALCULATION FOR THE ANCHOR

Verifying the stability:

Bending

$\Sigma \text{stabilizingM}/\Sigma \text{bendingM} > 2$

$\Sigma \text{stabilizingM} = P_1.X_1 + P_2.X_2 + P_3 \times X_3 + F_1.\cos 36° \times 1.20 + (F_3 + F_8)\cos 5° \times 1.20$
$+ (-F_2 + F_4)\sin 36° \times 1.20$
$= 6492.2 + 988.7 \times \cos 36 \times 1.20 + 381.53 \times \cos 5° \times 1.20 + 1059.02 \times \sin 36° \times 1.20$
$= 6492.20 + 959.85 + 456.09 - 275.78 = 7631.36$ kg-f.m

$\Sigma \text{bendingM} = -F_1 \times \sin 36° \times 0.80 - (F_3 + F_8)\sin 5° \times 0.80 + (-F_2 + F_4)\cos 36° \times 0.80$
$= -988.7 \times \sin 36° \times 0.80 - 381.53 \times \sin 5° \times 0.80 - 390.92 \times \cos 36° \times 0.80$
$= -464.91 - 26.60 - 253.0 = -744.51$ kg-f/m

Then: $7631.36/744.51 > 2$
$10.25 > 2$ ... OK

Sliding

$\Sigma F_x < \mu.\Sigma F_y$

$\Sigma F_x = -F_1 \times \sin 36° - (F_3 + F_8)\sin 5° + (-F_2 + F_4)\cos 36°$
$= -988.7 \times \sin 36° - 381.53 \times \sin 5° + 1059.02 \times \cos 36°$
$= -581.14 - 33.25 - 316.32 = -930.71$ kg-f

$\Sigma F_y = -P_1 - P_2 - P_3 - F_1.\cos 36° - (F_3 + F_8)\cos 5° + (F_2 - F_4)\sin 36°$
$= -6578 - 988.7 \times \cos 36° - 381.53 \times \cos 5° - 1059.02 \times \sin 6°$
$= -6578 - 799.87 - 380.07 + 230.23 = -7527.71$ kg-f

It was confirmed that: 930.71 kg-f $< 0.5 \times 7527.71$ kg-f
$930.71 < 3763.85$ ... OK

Soil stability

Calculation of R_y, and X (Figure A12.16)

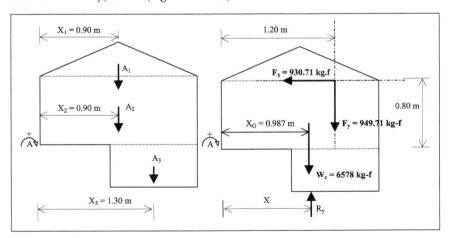

Figure A12.16 Diagram of partial weights of the concrete anchor and acting forces (Case II, when the pipe contracts)

Apply ΣM_A: $-R_y.X + 6578 \times 0.987 + 949.71 \times 1.20 - 930.71 \times 0.80 = 0$
$R_y.X = 6887.56$ kg-f.m
In the figure: $R_y = W_c + F_y$
$R_y = 6578 + 949.71 = 7527.71$ kg.f

$X = 6887.56/7527.71 = 0.914$ m

Calculation of the eccentricity factor $e = X - b/2 \rightarrow 1.055 - 1.80/2 = 0.156$ m

Calculation of the maximum and minimum bearing capacity of the soil

$$\sigma = \frac{R_y}{A}\left[1 \pm \frac{6 \times e}{b}\right] = \frac{7527.71}{180 \times 130}\left[1 \pm \frac{6 \times 0.014}{1.80}\right] = 0.32169\,[1 \pm 0.046]$$

a) $\sigma_{max} = 0.32169(1.046) = 0.34$ kg-f/cm^2
b) $\sigma_{min} = 0.32169(0.48) = 0.31$ kg-f/cm^2

It was confirmed that 1 kg-f/cm^2 > 0.34 kg-f/cm^2 > 0.31 kg-f/cm^2 ... OK

It was provided that in both cases, the anchor complies with the stability conditions, therefore the dimensions and shape were accepted as anchor 2 for the project.

Spreadsheet for calculating the anchor

Two cases are included in the attached sheets: an anchor for steel pipes and an anchor for PVC pipes.

In both cases, the spreadsheet contains two parts.

The first part is the information entered into the database in accordance with the profile and preliminary sizing of the anchor, design data, profile of the pipeline, physical and mechanical characteristics of the pipes and the anchor, etc.

The second part consists of the formulas entered in Excel language, which calculate the intervening forces, in accordance with the definition of each force. This also verifies the stability of the anchor to sliding and foundation soil conditions.

These spreadsheets are also applicable to the profiles of each mini and micro hydroelectric scheme, as reliable results were obtained in the projects implemented.

Should the profile be changed, the project designer would also have to verify whether it is necessary to change or modify the spreadsheet software.

The spreadsheets below correspond to two types of anchor in the Chetilla MHS. Anchor No. 02 located in the PVC section of the pipe faces out and anchor No. 04 located in the steel section of the pipe faces in. Both cases correspond to the same examples previously drawn up manually.

CALCULATION FOR THE ANCHOR 343

Table A12.2 Design of anchor no. 02 for the PVC pipe, Chetilla MHS

Description	Value	
H	69.00	H = static pressure head (m)
Q	0.08	Q = design flow (m³/s)
D_i	0.20	D_i = inner diameter of the pipe (m)
t	0.01	t = thickness of the pipe (m)
D_e	0.22	D_e = outer diameter of the pipe (m)
A	36.00	A = angle upstream from the horizontal
B	46.00	B = angle downstream from the horizontal
Pe_T	1450.00	Pe_T = specific weight of the pipe (kg-f/ m³)
Pe_A	1000.00	Pe_W = specific weight of the water (kg-f/ m³)
W_T	9.88	W_T = weight of the pipe (kg-f/m)
W_A	30.85	W_A = weight of the water (kg-f/m)
L_1	30.00	L_1 = average distance between the 1st and 2nd anchors (m)
L_2	55.00	L_2 = length of pipe subject to movement (m)
U	0.40	u = friction coefficient between the support and the pipe material
L_4	57.50	L_4 = length of the expansion joint with the anchor (m)
Sadm	1.00	admissible capacity of the soil (kg-f/cm2)
F1	988.67	
F2	725.02	
F3	377.98	
F4	334.00	
F5	0.00	
F6	0.00	
F7	0.00	
F8	3.55	
F9	0.00	
Data		
a	1.00	
b	1.80	
c	0.50	
d	0.80	
e	0.40	
f	1.30	

Table A12.3 Design of anchor no. 04 for the steel pipe, Chetilla MHS

Description	Value	
H	175.00	H = static pressure head (m)
Q	0.080	Q = design flow (m^3)
D_i	0.202	D_i = inner diameter of the pipe (m)
T	0.00817	t = thickness of the pipe (m)
D_e	0.219	D_e = outer diameter of the pipe (m)
A	32.000	α = angle upstream from the horizontal
B	0.000	β = angle downstream from the horizontal
ρ_t	7860.000	ρ_t = specific weight of the pipe (kg-f/m^3)
ρ_a	1000.000	ρ_a = specific weight of the water (kg-f/m^3)
W_t	44.182	W_t = weight of the pipe (kg-f/m)
W_a	32.047	W_a = weight of the water (kg-f/m)
L_1	51.000	L_1 = av. distance between 1st and 2nd anch. (m)
L_2	97.000	L_2 = length of pipe subject to movement (m)
U	0.5	u = friction coefficient between the steel and C°
L_4	99.50	L_4 = length of expansion joint to the anchor
Sbearing	1.00	admissible capacity of the foundation soil
F1	3296.95	
F2	3135.33	
F3	-3149.19	
F4	2329.57	
F5	0.00	
F6	2020.00	
F7	931.47	
F8	-10.81	
F9	0.00	
Data		
a	1.70	
b	2.30	
c	0.00	
d	1.00	
f	1.70	
g	1.80	
e	0.20	

EXAMPLE 13
Calculation of the foundation of a horizontal shaft Pelton turbine and generator (see Chapter 11)

The foundation for the horizontal shaft Pelton turbine and generator for the Sondor MHS was designed with the following data:
Q = 150 l/s, H_b = 131 m, P = 120 kW (electric)
Type of foundation soil: compact clay with 2" to 6" (5–15 cm) angular gravel, with an admissible bearing capacity of 2 kg/cm^2.

Table A13.1 Characteristics of the equipment supplied

Equipment	Supplier	Element	Weight (kg-f)	Dimensions
Horizontal shaft Pelton turbine with two jets (ITDG technology)	Peruvian energy technology, closed corporation TEPERSAC	Inlet pipe, elbows	50	8" × 4" (20 cm × 10 cm)
		Valves (02)	20	4" (10 cm)
		Casing	160	800 mm × 450 mm × 900 mm × 8 mm
		Shaft	55	Φ 80 mm × 850 mm
		Pulley	60	Φ 550 × 4 V-shaped channels
		Runner	80	700 mm
		Nozzles (02)	22	
		Bearers and bearings	18	
		Metal anchor base	42	Exterior: 95 mm × 60 mm, 75 mm wide, 8 mm thick
Three-phase, asynchronous generator, self-exciting, 220V, 60Hz 1800 RPM, 120 kW	SATURSA	Unit, including the pulley	540	Φ 570 mm × 920 mm
		Anchor base	38	800 mm × 450 mm × 75 mm

Sizing and calculation of the total static load

The sizing is shown in Figure A13.1.

346 DESIGNING MINI AND MICRO HYDROPOWER SCHEMES

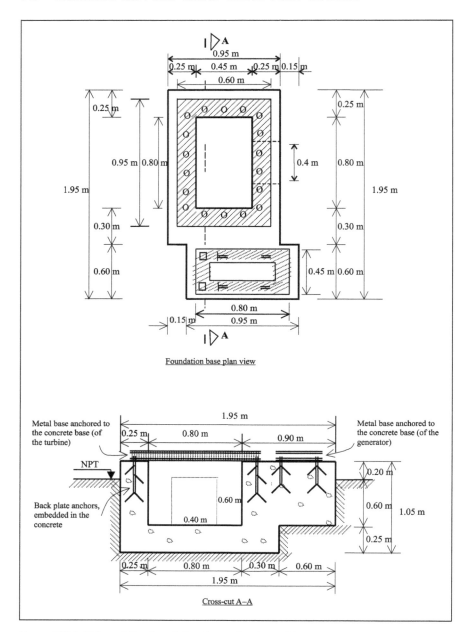

Figure A13.1 Sizing of the foundation bases for the turbine and generator

CALCULATION OF THE FOUNDATION OF A HORIZONTAL SHAFT TURBINE

Calculation of the total load

- Weight of the foundation base:
  ([1.95 × 0.95 × 1.05] – [0.60 × 0.95 × 0.25] – [0.45 × 0.80 × 0.80]) × 2400
  = (1.945 – 0.142 – 0.288) × 2400 = 3636.00 kg-f
- Weight of the equipment:
  Turbine: 507 kg-f
  Generator: 578 kg-f
- Additional weight equivalent to the spinning elements of the turbine and generator
  12(pulley + runner + shaft) + 12(0.5 × 540) = 12(60 + 80 + 55 + 270)
  = 5580.00 kg-f
- Live load:
  = 1.95 m × 0.95 m × 250 kg-f/m^2 = 463.12 kg-f
- Total load: 3,636 + 507 + 578 + 5580 + 463.12 = 10764.12 kg-f

Bearing capacity of the soil

10764.12 kg.f/195 cm × 95 cm = 0.58 kg/cm^2

Verification

Admissible capacity of the soil > Bearing capacity

2 kg/cm^2 > 0.58 kg-f/cm^2 ... OK

Characteristics of the reinforced concrete base for 120 kW of power

- f'_c = 210 kg-f/cm^2
- Vertical and horizontal steel, at least ½ inch diameter (f_y = 4,200 kg/cm^2)
- Thickness of the walls on the base of the turbine: 25 cm
- Steel distribution (Figure A13.2)

348 DESIGNING MINI AND MICRO HYDROPOWER SCHEMES

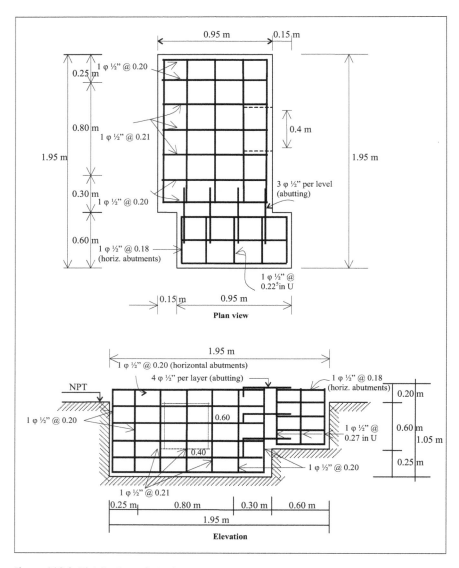

Figure A13.2 Distribution of steel structure

EXAMPLE 14
Calculation of the foundation for a vertical shaft Pelton turbine and generator (see Chapter 11)

The foundation for the turbine and generator in the Yanacancha MHS was designed with the following data:
Q = 150 l/s
H_b = 50 m
Power = 45 kW (electric)
$\sigma_{Tadmissible}$ = 1.5 kg-f/cm²

Table A14.1 Characteristics of the equipment supplied to the 'Yanacancha' MHS

Equipment	Supplier	Element	Weight (kg-f)	Dimensions
Vertical shaft Pelton turbine with three jets (ITDG technology)	Peruvian energy technology, closed corporation TEPERSAC	Inlet pipe, elbows	60	8" × 4"
		Valves (03)	24	4"
		Casing	140	1100mm × 1100 × 410mm × ¼"
		Shaft	30	Φ de 75 mm × 700 mm
		Pulley	42	Φ de 710 mm
		Runner	55	600 mm
		Nozzles (03)	15	
		Supports, roller bearings and bearing housings	25	
		Metal anchor base	—	
Three-phase, asynchronous generator, self-exciting, 220 V, 60 Hz, 1800 RPM, 50 kW	SATURSA	Unit, including pulley	350	Φ 430 mm × 600 mm
		U-shaped anchor base and support attached to the casing	62 26	800 mm × 740 mm × 10 mm 530 mm × 150 mm × 10 mm
		Anchored base		740 mm × 550 mm × 8 mm

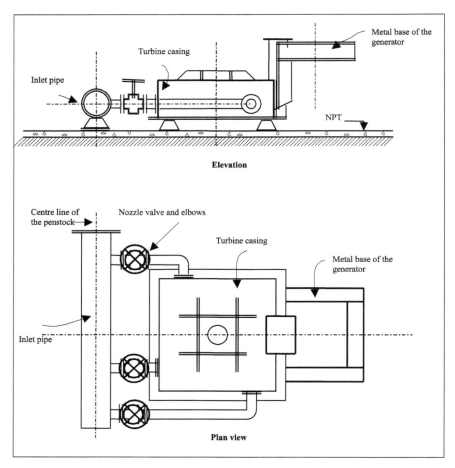

Figure A14.1 Pre-assembly of the vertical shaft Pelton turbine and the metal base of the generator

Sizing and calculation of the total static load

In accordance with the dimensions of the equipment and with a pre-assembly to scale in plan view and in elevation in the office, or in the powerhouse, (inlet pipe, casing and generator base) and taking the centre line of the penstock as a reference, the pre-assembly scheme shown in Figure A14.1 was determined.

Based on this pre-assembly, the dimensions of the reinforced concrete structure were determined, both in plan view and in elevation.

The foundation walls of the turbine are 0.25 m thick; the floor and the bed together are 0.30 m thick. The inner dimensions (pit) are determined by the inner measurements of the casing, with a slight variation to ease the flow of water from the turbine.

CALCULATION OF THE FOUNDATION FOR A VERTICAL SHAFT

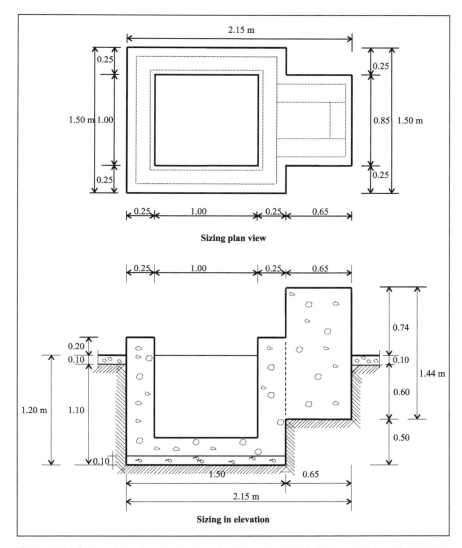

Figure A14.2 Sizing of the foundation base for the vertical shaft Pelton turbine and generator

The depth of the pit for the turbine foundation is determined by the thickness of the floor and the bed (0.30 m), plus the minimum height of the opening to the outflow channel, which is also an access underneath the casing (0.80 m), and by the thickness of the 0.30 m tall support bracket, the top of which is 20 cm above the finished floor.

The top level of the generator base is at a higher level than the turbine casing, because the generator pulley must be at the same level as the turbine pulley which is coupled to the top end of its shaft.

Calculation of the total load

- Weight of the foundation base
  ([1.50 × 1.50 × 1.30] − [1 × 1 × 1.10] + [0.65 × 0.85 × 1.44]) × 2400 + (0.10 × 1.50 × 1.50 × 2200)
  = 6289.44 + 495 = 6784.44 kg-f
- Weight of the equipment
  Turbine: 391 kg-f
  Generator: 438 kg-f
- Additional weight equivalent to the spinning elements of the turbine and generator
  12(pulley + runner + shaft) + 12(0.5 × 350) = 12(60 + 80 + 55 + 175)
  = 4440 kg-f
- Live load
  (1.50 m × 1.50 m + 0.65 m × 0.85 m) 250 kg-f/m² = 700.62 kg-f
- Total load
  6784.44 + 391 + 438 + 4440 + 700.62 = 12754.06 kg-f

Bearing capacity of the soil

12754.06 kg-f / (150 × 150 + 65 × 85)cm²
= 0.46 kg-f/cm²

Verification

Admissible capacity of the soil > Bearing capacity
1.5 kg/cm² > 0.46 kg-f/cm² … OK

Characteristics of the reinforced concrete base, for 45 kW of power

- f'_c = 175 kg-f/cm² (minimum)
- Vertical and horizontal steel, at least ½ inch (13 mm) diameter (f_y = 4,200 kg/cm²)
- Thickness of the walls of the turbine base: 25 cm
- Steel distribution (see Figure A14.3).

CALCULATION OF THE FOUNDATION FOR A VERTICAL SHAFT   353

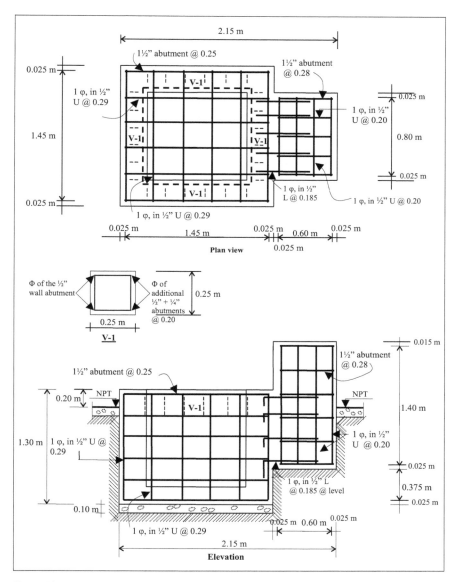

Figure A14.3 Steel structure in the foundation base for the vertical shaft Pelton turbine and generator
Note: ¼" = 6 mm; ½" = 13 mm; 1½" = 38 mm

EXAMPLE 15
Non-contact hydraulic seals for small micro and mini hydro turbines

The use of hydraulic seals for small hydro turbines is an innovation that reduces the operation and maintenance costs and improves the overall efficiency and performance of hydropower systems. A hydraulic seal is composed of two small discs installed in the shaft of the turbine to prevent water leaking out of the turbine casing. Hydraulic seals work on the same principle as disc pumps; they have no surfaces in contact and therefore have a long life (do not need replacement during the life of the turbine).

This short annexe provides a general view about why sealing is needed for hydraulic turbines, the use of conventional seals, the convenience of using hydraulic seals for micro and mini hydropower technologies; and the principles of hydraulic seals. And it provides a simple method to dimension the main parameters of the hydraulic seal. Hydraulic seals have been tested extensively in all Practical Action installations in Peru since year 2000 and a number of installations have also been carried out in other countries.

Conventional seals for hydraulic turbines

The conventional way of sealing water turbines is using mechanical seals. The basic layout of a mechanical seal is simple; it consists of a collar mounted on the shaft of the turbine which uses springs to push a ring (which also rotates with the shaft) against another ring which is held stationary. The rings rotate against each other riding on a thin layer of lubricant, and the springs hold them so tightly together that leakage through the seal is reduced to minimum (generally reduced to negligible amounts of water). The mating surfaces of the rings must be perfectly flat and extremely well finished, and are manufactured to precise tolerances. The rings must be made of hard materials to endure the pressure and wear, so they are usually made of ceramic, carbon, silicon carbide, tungsten carbide or similar materials. The stationary 'seat' is held in place and maintains a static seal with the mounting housing using gaskets or '0'-rings. The main problems of using conventional seals for hydraulic turbines are:

1. Short life span of the seal. Depending on the materials used, mechanical seals have relatively short life spans (a few years). Therefore a periodic investment is required to replace the seals. The replacement of the seals is generally budgeted as part of maintenance costs of the system.
2. Income losses due to stoppage time to change seals. Whenever the system has to stop, electric generation stops and consequently electricity sales cease.

3. Energy losses. Conventional seals generally imply very small losses of efficiency due to the friction between moving parts of the seal, which also imply small energy losses.

Contribution of this innovation

The use of hydraulic seals overcomes all the problems confronting conventional seals:
1. Long lifespan. Hydraulic seals do not have surfaces in contact; therefore there is no friction between surfaces and no wearing out.
2. Hydraulic seals are simple and can be manufactured at very low cost, and are easy to install and maintain.
3. Hydraulic seals generally require almost negligible energy to avoid water leakages, therefore negligible energy losses from sealing.

Working principles

Hydraulic seals operate on the principle of the boundary layer and viscous drag (Smith, 2001). For a disc pump, the discs create 'drag', and an initial 'boundary layer' of fluid adheres to their surfaces as flow enters the pump and then water is expelled. The main difference between a disc pump and a hydraulic seal is that while in the disc pump water reaches the centre of the discs through an inlet pipe, in the hydraulic seal water reaches the periphery of the discs in the form of a spray after impinging on the buckets/blades (of action turbines), and the work of the hydraulic seal is to stop the spray reaching the centre of the discs and the turbine's shaft, and therefore stopping water from exiting through the joint shaft-casing. (Hydraulic seals have been used very successfully with action turbines; however, more tests and field experience are still required to prove their convenience for reaction turbines.)

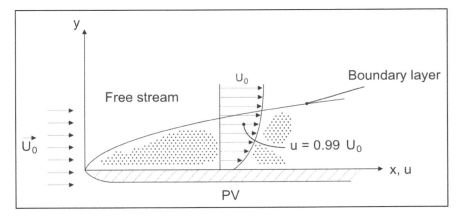

Figure A15.1. Boundary layer

According to the principles of the boundary layer (see Figure A15.1), when a fluid moves over a stationary surface, the fluid particles (the layer) in contact with the surface is stagnated, and the subsequent layers increase speed as their distance from the surface increases up to a value of U_o. The distance from the wall to the layer with a speed U_o is known as (δ).

Components of a hydraulic seal

The hydraulic seal has two main components:
1. A disc which is fixed to the casing of the turbine, called the 'stator disc' (see Figure A15.2) and
2. A rotating disc which is attached to the shaft of the turbine and rotates with it, called the 'rotor-disc'.

The stator disc is separated from the turbine casing by a small tubular sleeve which is long enough to either bolt or weld to the wall of the casing. The two components of the seals should be carefully manufactured in order to have uniform surfaces, and uniform separation between the discs. The rotor is fixed in the shaft of the turbine as shown in Figure A15.2.

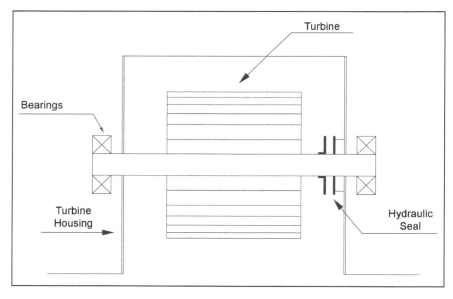

Figure A15.2 Position of the hydraulic seal

NON-CONTACT HYDRAULIC SEALS 357

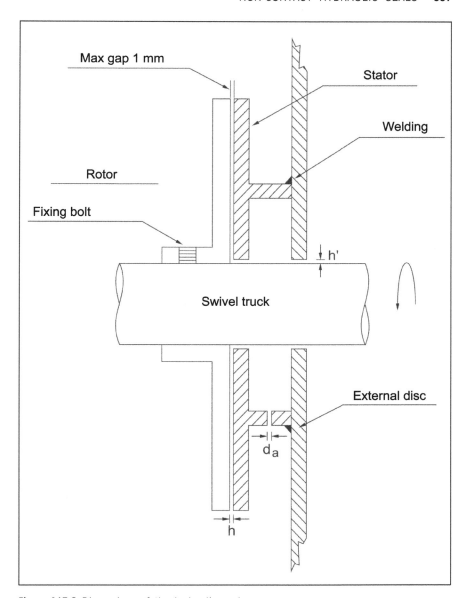

Figure A15.3 Dimensions of the hydraulic seal

Working conditions

Currently all the turbine manufacturers in Peru are using this technology for action turbines (Pelton, cross flow, Turgo). There is not much experience in using this technology for reaction turbines (propeller, Francis turbines), the main reason being that in action turbines water reaches the hydraulic seals at atmospheric pressure at the exit of the buckets or blades, while in reaction turbines water reaches the joint shaft-casing at much higher pressures than atmospheric, and therefore the seals need to be designed to pump water at even higher pressures. Once the hydraulic seals are installed these can operate under any of the following working conditions

1. *When the turbine is operating.* In action turbines, water reaches the hydraulic seals as a spray; and the spray will impact over the components of the seal and turbine's shaft. As soon as any water reaches the discs of the seals it will be expelled into the turbine casing, preventing leakages outside the turbine casing.
2. *When the turbine is not in operation.* Two situations can happen when the turbine is not operating: (a) the gate valves are completely closed, in which case there is no water going through the joint shaft-casing and therefore no risk of leakage; (b) the gate valves are partially open but the turbine is not operating; in which case the hydraulic seal is not rotating, and therefore water can reach the joint shaft-casing and can eventually exit the casing. To avoid this, the small sleeve of the stator disc has a hole to allow water to exit by gravity (see Figure A15.3)
3. *Flooding of the casing.* This situation is very unlikely to happen, but if it were to occur, the way to avoid water leakages is by making the hole sufficiently large to allow the exit of all possible water that gets through the discs and the shaft-disc joints (see sizing recommendations below).

Sizing recommendations for the hydraulic seal components

The following recommendations for dimensioning the seal components are the result of analysis of continuity, thickness of the boundary layer and separation of the discs and field trials in actual installations.

Parameters
h, separation between the stator and rotor
h', gap between the turbine shaft and its casing
D, diameter of the seal (rotor and stator discs)
D_e, diameter of the cylinder (separation of the stator from the turbine casing)
d, diameter of the shaft of the turbine
D_a, diameter of the drain hole

Size
0.8mm<h<1.0mm
h'<0.2mm
2d<D<3d
6h< D_a <10h
1.25d< D_e <1.5d

Bibliography

Discflo, 'The future of pump Technology', Disc Pumps, SP Series of sanitary pumps, 1817 John Towers Av, El Cajon, CA, USA.

Schutz, Gunter et al., (1997) *Friction vacuum pumps with pump sections of different designs*, USA Patent, December.

Murata, Susumu et al. (1976) 'A study of a disc friction pump (second report, experiments on flow through Corotating discs and pumps performance)', *Bulletin of the Japan Society of Mechanical Engineers* Vol. 19, No. 136.

Robanoff, Val and Robert R. Ross (undated) 'Centrifugal pump design and application, A practical reference stressing hydraulic design and performance prediction, analysis and evaluation', web: http://books.google.co.uk/books?id=-LPJ8yN8OSEC&printsec=frontcover&dq=friction+disc+pumps&source=gbs_similarbooks_s&cad=1

Sanchez, Teodoro (2009) 'Sellos sin contact para turbinas hidraulicas', article presented at the 'The Latin American Network Bi-Annual Meeting', held in Cajamarca, Peru, 2009.

Smith, Jeannie (2001) 'Disc pumps keep fluids moving', Paint and Coatings Industry, available at http://www.pcimag.com/Articles/Feature_Article/2db27716466a7010VgnVCM100000f932a8c0

Universidad de Santiago de Chile, 'Estudio de Desarrollo de Capa Límite', Facultad de Ingeniería, Departamento de Ingeniería Mecánica, Asignatura Mecanica De Fluidos Ii, Nivel 03, Experiencia C902.

Milton Keynes UK
Ingram Content Group UK Ltd.
UKHW021044020524
442115UK00006B/262